T0220581

FIGURING IT OUT

NUNO CRATO

FIGURING IT OUT

ENTERTAINING ENCOUNTERS WITH EVERYDAY MATH

 Springer

Nuno Crato
Universidade Técnica de Lisboa
Inst. Superior de Economia e Gestão
Dept. Matemática
Rua Miguel Lupi 20
1200 Lisboa
Portugal
ncrato@iseg.utl.pt

Original Portuguese edition published by Gradiva Publicações lda., Lisboa, Portugal, 2008 original title: A Matemática das Coisas

ISBN 978-3-662-50552-6   ISBN 978-3-642-04833-3 (eBook)
DOI 10.1007/978-3-642-04833-3
Springer Heidelberg Dordrecht London New York

Cover design: eStudio Calamar S.L.

Printed on acid-free paper

Springer is part of Springer Science+Business Media (www.springer.com)

PREFACE

"When I tell people that I am a mathematician, they jokingly ask if I could help them balance their bank account. Then, when I tell them I make lots of counting mistakes, they think I must be a pretty mediocre mathematician."

That is what one mathematician friend of mine once told me, but it could as easily have come from just about any mathematician, as almost everyone in this field complains of how misunderstood the profession is. There really are a lot of people who have no idea what it is that mathematicians do.

Math, of course, is an integral part of our daily lives. The 20th century could not have been the most revolutionary one hundred years in the history of science, as indeed it was, without the extraordinary advances that took place in the field of mathematics. Computers could not have been created without binary logic, group theory and the mathematical concept of information. Telephones would not work if mathematicians hadn't developed the statistical study of signals and the algorithms to digitalize and compress data. Automated traffic lights would no doubt effect chaos, rather than order, if advances in a field of mathematics called Operations Research had not occurred.

But despite its crucial importance, mathematics is frequently viewed as an insular, even irrelevant field into which few interesting people venture, and which has little to contribute to our daily lives. Even the well-educated often demonstrate a surprising ignorance of the history of mathematics and its advancements.

I would venture that if you asked an intellectual to name two or three renowned 20th century philosophers, there would not be many who could not respond without hesitation. I would also say that most reasonably educated people could easily name two or three great contemporary composers. Many of them would also have little difficulty in identifying half a dozen modern schools of art, from cubism to minimalism. But mathematicians and fields of mathematics? Few people know who David Hilbert was or what the formalist school was, or the important part that Andrey Kolmogorov and John von Neumann played with respect to probability studies.

This book is full of stories about math, with few equations, lots of examples and many applications. Math is a fascinating science, of fundamental importance for our history and always present in our daily lives. Many things would not be possible without math: Picasso's art, online bank transactions, house numbers and A4 paper sizes, modern maps and the defeat of Hitler. Math applications appear where you would least expect them. The history of math is the history of winners.

Lisboa, Portugal   Nuno Crato

CONTENTS

EVERYDAY MATTERS

The Dinner Table Algorithm

If you want to invite some friends to a dinner party, but your dining table will only accommodate four people, then you might be faced with a dilemma: how do you choose three compatible dinner companions from among your five closest friends? Your buddy Art has recently broken up with his girlfriend Betty, who is now dating Charlie. Charlie and Art have managed to remain friends, but Charlie is not speaking to Dan, who won't go anywhere without Eva, who can't stand Art. So how can you choose your three dinner companions to have a pleasant, hassle-free evening? The best way, believe it or not, would be to make use of an algorithm, which is a set of rules that enable you to search systematically for an answer.

Algorithms are much loved by mathematicians as well as computer scientists. Even though some algorithms are very complex, the simplest can sometimes be the most effective. In our case we can follow a systematic process of trial and error, which may be quite an efficient algorithm, despite its apparent simplicity.

So let us start by choosing your friend A. Under the circumstances, we immediately see that you cannot possibly also invite your friend B. You could invite C, but he wouldn't come unless B was also invited. And so it goes on. It seems there is no scenario under which A could be included, which means we need to start again, this time with B, and keep going until we have found three companionable friends for the dinner party. Will that be possible in this case? Or will we have to give up, forced to admit that human relationships are more complicated than algorithms?

N. Crato, *Figuring It Out*, DOI 10.1007/978-3-642-04833-3_1,
© Springer-Verlag Berlin Heidelberg 2010

This type of problem is known as a satisfiability problem. Mathematicians call them SAT problems, which keeps things simpler. The dinner party mentioned above is an example of a "2-SAT" problem, as each restriction contains two variables ("A or B", "A and C", etc.). The problem would become more complicated if Art, Charlie and Dan were inseparable, i.e., if we had to take three variables in each restriction into account ("A and B and C" or "A and C or D").

Such problems are known as "3-SAT" problems. And it is also possible to imagine restrictions of a more general type, which give rise to "k-SAT" problems.

Although this example may seem trivial, a similar approach can be applied to many basic tasks, such as drawing up timetables in large schools, organizing conferences, or planning flight schedules for airlines. It is the basis of a new branch of mathematics called computational complexity, which aims to study and classify problems in terms of their inherent difficulty. When such SAT problems could only be solved manually, one after the other, it was difficult to study many of their characteristics. However, from the time it became possible to employ computers to solve them, attempts have been made to study the complexity of the processes used to solve them, i.e., the algorithms, and to evaluate the time that it takes a computer to solve them.

In 1959, Richard Karp was still a 24-year-old mathematician who had just earned a Ph.D. from Harvard and begun to work at the IBM research laboratory at Yorktown Heights, NY. At the time, computers were in their infancy, but the invention of transistors made it possible to incorporate more and more elaborately designed circuits. Karp's task at IBM was to find an automatic process for designing circuits with as few transistors as possible. Written as a computer program, the algorithm he wrote was limited to checking out all the possible circuits and calculating their costs. Later, in 1985, when he was presented with the prestigious Turing Award given by the Association for Computing Machinery, Karp recalled that although this approach seemed simple, it contained a basic problem: "The number of circuits that the program had to comb through grew at a furious rate as the number of input variables increased, and, as a consequence, we could never progress beyond the solution of toy problems."[1] Karp, who spent 10 years more at IBM before becoming a professor at Caltech, had identified a phenomenon that came to occupy the attention of hundreds of researchers and to generate thousands of studies: the problems might well be simple and the technique might be easy to apply, but they could rapidly grow to become impossible to solve, even when using the most powerful computers. Mathematicians, logicians and computer scientists spent many years subsequently trying to devise more efficient algorithms, but always arrived at the same result: there are problems that can be resolved simply and that have a complexity that increases in a controlled fashion, and there are problems that quickly become impossible to solve because their complexity increases exponentially with the number of variables and restrictions.

At present, a distinction is made between the "type P" problems, in which the complexity increases in polynomial time with the rise in the number of variables, and the "non-P" type problems, in which this does not happen. In particular, there is a class of non-P problems that are all reducible to each other and whose solution can be checked in polynomial time. These are the so-called "NP-complete" problems (nondeterministic polynomial). Even though solutions for these problems can be checked efficiently, to find such solutions there are known algorithms that increase dramatically in computing time (more than polynomially) as their dimension grows. These problems thus become impractical when the number of variables increases. It is still not known if type NP-complete problems are amenable to a type P approach. This question was also posed by Karp in 1985 during his Turing Award speech, but even today remains a major unsolved issue in computer science. Specialists assume that these are two different and irreducible types of problems, but they have not been able to prove this yet.

Our dinner table dilemma, which is a 2-SAT problem, is of type P. Even if we had to select thirty persons from a group of 50 instead of having to choose three of our five friends, a computer program could find

---

[1] From Karp's 1985 Turing Award lecture "Combinatorics, complexity, and randomness" (in http://awards.acm.org)

a solution rapidly or indicate that there is no possible solution, which would be equally important to know.

And if we were, say, holding an event at the UN and had to select 300 persons from a list of 500 possible guests, this would indeed keep the computer busy for a little longer, but we would still have an answer in a reasonable amount of time.

Strangely enough, though, we enter another world entirely when we move on to a 3-SAT problem by inserting restrictions such as "either not including Art and Betty or Charlie". We then cross the line dividing type P problems, for which we will eventually find a solution, from NP-complete problems, when having a few dozen friends is enough to make it impossible for any computer in the world to organize our dinner table in time.

CUTTING THE CHRISTMAS CAKE

When a small cake has to be cut in two pieces to be shared by two people, and the person who cuts the cake is also the person who chooses which half to take, then there is no guarantee that one of the two people will not be disadvantaged. The best way to avoid any complaints about the division of the cake is for one person to cut the cake and the other to choose which half to take. This way, it is in the first person's interest to divide the cake as fairly as possible, as otherwise he or she might very well end up with the smaller piece. It is a wise solution, requiring that two persons, basically motivated by egotism, cooperate with one another in such a way that neither is deprived of a fair share.

This well-known anecdote is applicable to many situations in our day-to-day lives, and not only ones involving cakes. However, the problem becomes more difficult when the cake has to be divided among more than two persons. How would you divide a cake among three people, for instance? Two cut and one chooses? Couldn't two of them conspire to deprive the third of a fair share? And what if many more people wanted a slice of the action? What if a cake had to be shared by twenty equally sweet-toothed persons?

That is not a trivial problem, and mathematicians are beginning to develop algorithms for equal shares. These algorithms can be applied in very diverse areas, ranging from personal matters like the sharing of an inheritance to affairs of state such as establishing international borders.

The "one cuts, the other chooses" algorithm can be directly applied to some situations in which more than two people are involved. If four people want a slice of the cake, for example, the algorithm is applied in

N. Crato, *Figuring It Out*, DOI 10.1007/978-3-642-04833-3_2,
© Springer-Verlag Berlin Heidelberg 2010

two steps. We start by assigning the four people to two groups, each containing two members. One of the groups then cuts the cake in two, and the other group chooses its half. In the second step each group divides its half of the cake, using the established procedure of one person cutting and the other choosing.

It is easy to see that this repetitive method can also work well with eight participants or indeed with any number that is a power of two. It is not so simple to find a solution when three persons want to share the cake. But if you think carefully about it, you will find a solution to this problem. Can you suggest a way?

However, mathematicians don't like methods that work only in special cases; they prefer to devise algorithms that can be more widely applied. The ideal would be to find methods that could be applied to any number of persons. One of these methods, first proposed by the Polish mathematicians Stefan Banach (1892–1945) and Bronislaw Knaster (1893–1980), resolves the problem utilizing what has been dubbed the "moving knife procedure". It is easier to explain if we take a loaf cake as an example.

The persons who want a slice of the cake gather round it while one of them begins to slide the knife along the cake. The knife keeps moving until one of the participants says "Stop!". At this precise moment the knife stops moving and a slice is cut from the cake and handed to the person who said "Stop". This person then has a slice that he or she considers to be at least a fair share of the cake – if he or she had thought that the knife had not yet traveled far enough to provide a fair share, then he or she would have remained silent. Now the others also had the chance to say "Stop", but they did not do so. So presumably they did not consider that the slice of cake offered was larger than a fair share – otherwise they would have claimed this slice.

After being given a slice, the first participant leaves the game while the knife continues to travel along the cake until one of the remaining participants says "Stop!" and is given the corresponding slice. This process is repeated until only two participants remain in the game. At this stage, the first person to speak receives the slice that is cut and the other receives the remainder of the cake.

The moving knife method can be used to divide a homogeneous cake into equal portions for an arbitrary number of persons. One person moves a knife along the cake until one of the participants says "Stop!" and claims the slice of cake that is cut at that point. The procedure is continued until another participant claims a slice, and so on until the cake has been divided into slices for each person

The interesting thing about this method is that, even considering the fallibility of each of the participants in assessing the right moment to say "Stop", none of them can claim that he or she has been disadvantaged. If any person has not in fact received their fair share, then it is their own fault, as he or she did not speak up at the right time.

This method seems to be perfect, but it fails to take some interesting aspects into consideration. It works well with a homogeneous cake, but would it work with a cake that has various ingredients that are distributed irregularly, like a Christmas cake? Would it be possible to devise an algorithm that guarantees that each person ends up with an equal quantity of glacé cherries, almonds, sultanas and dough? An answer to this question is provided by a theorem the Polish mathematician Hugo Steinhaus (1887–1972) proved in the 1940's and that came to be known by the curious name of the "ham sandwich theorem". Let us take a three-dimensional object with three components such as a sandwich consisting of bread, butter and ham – it does not matter if these components are distributed equally or not, are concentrated in different areas or are spread uniformly. What this theorem proves is that there is always a plane that divides the object in two halves in such a way that each half contains an equal quantity of the three components. In other words, even if the ham or the butter are distributed unequally, there is always a way to cut the sandwich into two completely equal halves.

In the case of a two-dimensional object an equal division works only with two components. Let us suppose that salt and pepper are spread on a table, for example. Steinhaus's theorem proves that there is always a straight line that divides the surface of the table into two sections containing equal quantities of salt and pepper. If there were three ingredients, let us say salt, pepper and sugar, it is easy to imagine a concentration of the substances in three different places so that it would be impossible to draw a straight line that would divide them equally. Generally the theorem states that for n dimensions there is always a hyperplane that simultaneously divides n components equally. As it seems that we live in a three-dimensional world, and as the Christmas cake has many more ingredients than just three, we have just learned that no knife exists that can cut slices of Christmas cake containing equal quantities of all the ingredients.

ORANGES AND COMPUTERS

For more than 2000 years mathematics has been making progress by means of rigorous proofs, based on explicit assumptions and logical arguments. The arguments should be faultless. But how can their validity be checked? This has always been the subject of debate and has never been completely resolved. The issue was rekindled at the end of the 20th century, when some prestigious mathematical journals accepted proofs completed with the help of computers. Should these proofs be accepted as legitimate? Should they even be considered mathematical proofs?

One of these disputes involved a well-known and easily understood problem: what is the best way to stack spheres? Is it the way that supermarkets sometimes stack oranges, in little pyramids structured in layers, with each orange sitting in the space between those on the layer below? This system seems more efficient than piling one orange exactly on top of another, for instance. But aren't there other more efficient ways to stack them?

Legend has it that this particular mathematical problem originated in a question that the English explorer Sir Walter Raleigh (1552–1618) posed to the scientist Thomas Harriot (1560–1621). Raleigh was interested in finding a procedure for rapidly estimating the quantity of his munitions. For this purpose, he wanted to be able to calculate the number of cannonballs in each pile simply by inspecting it, without having to count them. Harriot was able to provide him with a correct and simple answer for square pyramidal piles: if each side of the bottom layer of the pile has k cannonballs, then the stack consists of $k(1 + k)(1 + 2\,k)/6$ cannonballs. So, for instance, if the bottom layer of a square pyramidal

N. Crato, *Figuring It Out*, DOI 10.1007/978-3-642-04833-3_3,
© Springer-Verlag Berlin Heidelberg 2010

pile has four balls on each side, then the pile has a total of 30 balls. You could check this yourself by stacking thirty oranges of your own.

Harriot studied various ways of stacking balls. Years later, he brought up the problem in a discussion with the German astronomer Johannes Kepler (1571–1630), who posed an even more interesting question: what is the most efficient way of packing spheres?

Kepler conjectured that the best way would be to put balls in parallel layers, with each layer disposed along a hexagonal grid. Balls on layers below and above should be inserted on the spaces formed by the balls on the other layers. Kepler concluded that there was no better solution than this one but he was unable to prove it mathematically. Centuries passed, and the problem became known as the sphere-packing problem. The astronomer's supposition became known as the Kepler conjecture. It was always admitted that the supposition was true, but nobody ever succeeded in proving it with absolute certainty.

Then in 1998, Thomas C. Hales, a mathematics professor at the University of Michigan, surprised the scientific community by providing a proof. After this, Kepler's conjecture seemed to have ceased being a simple hypothesis and to have become a perfectly proven theorem. However, there was a problem with all this. Just one minor problem... the proof had been derived with the help of a computer.

Hales had explicitly resolved many of the steps that were required to prove the hypothesis, but he had left others to be tested automatically using software specially written for this purpose. He claimed that combining the results from the computer with his own work would unquestionably prove the theorem. This was not the first time that a proof had been made with the assistance of a computer. In 1976, Wolfgang Haken and Kenneth Appel, from the University of Illinois, had also used a computer to attain another of the great goals of mathematics – the proof of the four colors theorem, which posits that four colors are sufficient to color a flat map in such a way that no two adjacent regions have the same color. And in 1996 Larry Wos and William McCune, of the Argonne Laboratory in the USA, used logical software to provide proof of another famous supposition, the "Robbins conjecture", a deep statement in mathematical logic.

As soon as Hales announced his achievement, the *Annals of Mathematics*, a prestigious scientific journal, offered to publish his work, but as is usual in academic circles, only after it had been peer-reviewed, that is reviewed by fellow experts. It then took years of work before a panel of 12 experts declared that they had been defeated by the enormity of the task. They confirmed that they were 99% certain that the proof was valid, but they could not succeed in independently verifying all the steps the computer had performed. The editor of the journal regretfully wrote to Hales that while the experts had approached their task with unprecedented vigor, they had become completely exhausted before being able to complete the verification.

The editors of *Annals of Mathematics* did eventually decide to accept the work performed by Hales, though they would only publish those parts that had been verified via explicit logical reasoning, as is normal in the field of mathematics. The computational parts of Hales' proof were published in another, more specialized journal, *Discrete and Computational Geometry*. The provision of computer-generated proof has thus been implicitly admitted into the realm of pure mathematics, but it continues to be regarded with suspicion. Will this ever change?

WHEN TWO AND TWO DON'T MAKE FOUR

Two and two always makes four. But the *four* can result from the sum "two plus two" or from the sum "one plus three". It would seem impossible to differentiate between the two fours. However, this problem has a tremendous practical importance for statistics.

In 1919, two American political scientists, William Ogburn and Inez Goltra, published a study on the voting behavior of Oregon women who had recently registered to vote for the first time. The two investigators only knew the total number of votes cast in the election, but had no information on voting patterns according to gender. "Even though the method of voting makes it impossible to count women's votes" they wrote, "one wonders if there is not some indirect method of solving the problem".[1] They decided to estimate the correlation between the number of votes cast in each district with the number of women who had voted in that district. In this way, in the districts with more women, they could attribute the departures from the mean to the higher number of women voters. Still, as the investigators themselves conceded, their method was fallible, as there could have been another explanation: men could have changed their voting habits in those districts that had a greater number of women.

The problem of reconstructing individual behavior from aggregate data came to be known as the *ecological inference problem* (as ecology is the science that is concerned with the relationships between the

---

[1] W. F. Ogburn and I. Goltra, *Political Science Quarterly* **34**, 413–433, 1919.

N. Crato, *Figuring It Out*, DOI 10.1007/978-3-642-04833-3_4,
© Springer-Verlag Berlin Heidelberg 2010

elements and their environment), but very few basic steps were taken to solve it.

Thirty years later an American sociologist named William Robinson published a study that decisively influenced the future methodology of the social sciences. Essentially, Robinson showed that the existing methods at that time did not permit the reconstruction of partial data from aggregate data, and he afterwards coined the expression "ecological fallacy" to describe the faulty inferences that could be drawn as a result. Robinson's study cast doubt on several strands of sociological investigation. Geopolitical studies, which were flourishing in France, Germany and the USA, practically ground to a halt when the validity of the methods then used was questioned.

However, the ecological inference problem is still a pressing question in applied statistics. The questions posed by the studies are too important for scientists to simply accept that no solution exists. The prime example that is usually cited is the attempt to understand the political and electoral success of the Nazi party in the early 1930s. In this case it is necessary to differentiate between the groups and classes that supported Hitler's rise to power. The sociologists have based such studies on the data for each electoral district, for which only aggregate data is available. They have no other option.

Another prime example of the importance of ecological inference is taken from epidemiology. The total number of persons affected during an outbreak of disease is often known, but the specific areas of the population that are most affected are often much less evident. The data are aggregated in the hospitals, but in less developed countries it is always very difficult to process them so that the zones where the epidemic is spreading most rapidly can be pinpointed quickly. An efficient method for comparing aggregate data with the existing parceled information (for instance, in some better-organized health centers) could be used to detect the origin of the epidemic and to help save many human lives.

Yet another example comes from marketing. The success or failure of an advertising campaign in attracting new customers can usually be measured, as can the age and income distribution of the target population. Nevertheless, it may be too costly to carry out the research that could pinpoint the age groups and social groups that showed the best responses to the campaign, so this effort is often not made, even though the resulting knowledge could provide essential data.

The methods used to date for ecologic inference have not been very successful, and at times they have even been disastrous. People usually cite ridiculous examples, such as a study carried out by a group of Israeli sociologists to forecast the number of voters who would remain loyal to the Labor Party, which resulted in a *negative number of voters!* Or the example of a US opinion-polling company that concluded that 120% of blacks in Louisiana would vote for the Democrats!

Gary King, a statistician and political scientist at Harvard, has succeeded in finding new solutions to the ecological inference problem.

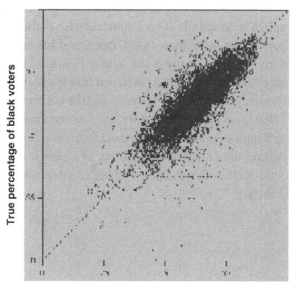

Estimated proportion of black voters

Comparing his estimates with results obtained at a later date, Gary King found a remarkable fit with the actual outcome. In the diagram the 3262 electoral districts in Louisiana are represented by a number of dots proportional to the number of voters in each district. Almost all the elements are located along the diagonal line, indicating that the estimated fraction and the actual fraction of black voters are practically identical

His method is much more complex than the normal multiple-variable procedures, as it is non-linear. The algorithm starts by analyzing the smallest units that can possibly be obtained.

This data is used to calculate the logical limits for each subgroup. For example, if a thousand voters voted for a certain candidate, the number of women who supported this candidate cannot be less than zero or greater than one thousand. These limits, which may seem trivial, introduce non-linearities into the statistical tools. The next step in the algorithm is to estimate a more probable value that maximizes the correlation of the estimated figures for each subgroup with the fragmentary parceled data that exists on some of the subgroups. Finally, these estimated figures are compared with the known values for some subgroups and then corrected.

The method is obviously quite complex, and a whole book was needed to explain it in detail.[2] The important fact is that Gary King tested the algorithm in more than 16,000 cases, and his estimates were shown to be remarkably close to the actual figures. The American Political Science Association (APSA) awarded him the Gosnell Prize for the "best methodological study" of the year, and the US National Science Foundation (NSF) was equally laudatory. Its director Frank Scioli said "I expect Gary King's solution will contribute to the production of more accurate, insightful data analysis in a variety of research studies, leading to more informed policy-making and better understanding of our economy and society, "[3]

---

[2] Gary King published his study in *A Solution to the Ecological Inference Problem*, Princeton University Press, 1997. The author has also made computer programs that permit the application of his method available on the internet. These programs, which run in the Windows environment and in the GAUSS language, are available free of charge at http://gking.harvard.edu.

[3] Quotation from http://www.nsf.gov/news/news_summ.jsp?cntn_id=102784

GETTING MORE INTELLIGENT EVERY DAY

Ever since intelligence tests were first invented, almost one hundred years ago, there has been a spectacular upsurge in their average results. The increase has been most surprising in the less specific tests, such as the intelligence quotient (IQ) tests, which are supposed to assess various types of intelligence. Could this be true, or are there major errors in the test concepts? This is a difficult question, and psychologists, statisticians and psychometricians do not agree on how to interpret the test results.

One thing seems to be certain: the increase in IQ test results is neither an isolated phenomenon nor a statistical artifact. The tests have been calibrated over the course of many years and have been taken by millions and millions of people all over the world, and everywhere the same phenomenon is observed: average persons who are tested today using old tests achieve results that would have been classified as highly intelligent just a few decades ago.

The arguments about what intelligence tests measure are as old as the tests themselves. Nevertheless, there is not a single scientist who today asserts that IQ tests are meaningless, just as there is no specialist who believes that they are infallible.

Intelligence tests, which establish a coefficient internationally known as IQ, attempt to measure various components such as memory, reading skills, spatial visualization, and arithmetical and logical capabilities. For a long time, in accordance with the statistical studies performed by Charles Spearman, it was accepted that these various components were closely correlated, and that there was a "general g factor" that underlay all the measurements. The IQ measurement was worked out

N. Crato, *Figuring It Out*, DOI 10.1007/978-3-642-04833-3_5,
© Springer-Verlag Berlin Heidelberg 2010

on the basis of a variety of tests. It was thought that the inherent errors in the different types of tests would even out the results, and the mean would indicate the g factor that was quantified in the IQ.

The theory of a single intelligence and of a "general g factor" has come to be viewed with increasing skepticism, however, particularly after the publication of the studies by Howard Gardner. The alternative view that he put forward holds that there are *different types of intelligences,* what he called *multiple intelligences.* A person with great mathematical and logical capabilities, for example, could still be deficient in the area of intuitive interpretation, just as a person with great powers of spatial visualization might have difficulty understanding elementary algebra. However, Gardner's followers have also been criticized for ignoring the correlation between the different aspects of intelligence, and the view that is prevalent today is more balanced.

Whether a g factor exists or not, the reality is that standardized tests combine a battery of partial tests that are weighted to produce an overall measurement. This measurement is then compared with the measured results attained by a great number of individuals within the same age group, and is then standardized so that the mean is 100 and the standard deviation is 15 points. In this way the mean intelligence for each age group at each period of time is always set at 100. The standardization, based on a Gauss curve, ensures that 95% of individuals have an IQ between 70 and 130. A value below 70 is considered to show mental deficiency, and a value above 130 to show exceptional intelligence.

The gain in IQ values is apparent when individuals take tests from bygone days. For example, it has been calculated that American children today have a mean IQ of 120 when assessed by the criteria used in 1932, which means that about 25% of them would be considered exceptionally intelligent, compared with only 2.25% in 1932.

How can this development be explained? Scientists offer different opinions. One of the most common explanations is that test-taking strategies have been learned and perfected over time. It seems obvious that today's young people and adults are much more familiar with standardized tests than those who took them at the beginning of the 20th century. Both scholastic standards and familiarity with

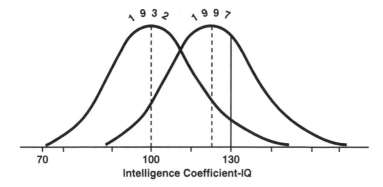

When a sample of American children were subjected to the Stanford-Binet intelligence test in 1932, the results showed a good fit with a Gauss curve with a mean of 100 and a standard deviation of 15. Only 2.25% of the children scored over 130. When children of the same age took the same test in recent years and the results were calibrated with the weighting factors used in 1932, the mean was 120, with 25% of the children scoring more than 130, so they would be considered exceptionally intelligent. Is that possible?

multiple-choice tests have increased tremendously, which is why this explanation appears very plausible.

While these factors do have some influence, the problem is that they only explain a small part of the phenomenon. On the one hand, an increase in scholastic standards would imply that the tests that have shown the greatest gains would be those most closely related to learning themes. But that is not the case. Rather, it is the more basic tests, such as those based on Raven's Progressive Matrices, that have resulted in the highest increases. On the other hand, comparative studies of young people exposed to different levels of scholastic standards show that scholastic standards can only explain a small part of the phenomenon, which means that this factor is not solely responsible for the huge gains that have been recorded over the past 100 years.

Nor does the explanation involving increased familiarity with standardized testing seem to hold water. Only a slight difference can be discerned when the results obtained by intensively coached children are compared with those of their less well-prepared counterparts. This difference in the mean results is limited to 5 or 6 points. Significant differences are not achieved even when young people and adults take *the same test multiple times*. This fact provides evidence of the consistency of this type of assessment, and reassures researchers that whatever it is that IQ tests are measuring, they are measuring it well.

The psychometrician Richard Lynn, from the University of Ulster in Northern Ireland, has documented that it is improvements in nutrition and hygiene that are largely responsible for vast improvements in our physical and mental health. Lynn goes on to argue that this has in turn produced an increase in mean stature, brain size and intellectual capacity.

Although this explanation seems credible, it too has been the object of counter-attacks by a great many psychologists and statisticians, who argue that such an explanation would imply a considerable increase in the intellectual capacities of one generation compared to the previous one, and that the facts do not support this. For example, nobody asserts that the average person today is a genius compared to the average individual at the beginning of the last century. The improvement seems to be limited to a certain type of abstract intelligence shown by testing.

The American psychologist and psychometrician Ulric Neisser, from Cornell University, has offered a simple and convincing explanation for this phenomenon. Neisser studied the types of tests in which the gains were most remarkable, and noted that the tests using Raven's Progressive Matrices were primarily responsible for the IQ gains. Well, these tests measure the capacity for abstract reasoning as well as for interpreting diagrams. Neisser argues that the 20th century saw a real explosion in the field of audiovisual media, from street advertising to films, cartoons and computer games. Neisser concluded that there are different forms of intelligence that develop better according to different types of experience. In fact, he said, we are much more expert than our grandparents with regard to visual analysis, but not with regard to other forms of intelligence.

THE OTHER LANE ALWAYS GOES FASTER

Jack and Anna leave their respective homes at 8 and have to drive over a bridge to get to the office where they both work. The traffic begins to back up long before they reach the bridge, but each of them handles the situation differently. Whereas Anna remains calm and stays in the right-hand lane, Jack, whose car is behind hers, soon switches over to the left-hand lane and overtakes her. Up ahead, his lane comes to a halt, and Jack is forced to sit there and watch as the cars in the right-hand lane now pass him. Then, taking advantage of a gap, he abruptly decides to rejoin the right-hand lane. A bad decision, as right then his new lane stops again. He waits a short time, frustrated and unable to do anything, until once more a gap opens up in the other lane. He makes use of it to change lanes again in an even riskier maneuver than last time. Now, he feels as if he is gaining ground, until the traffic grinds to a halt yet again. This pattern is repeated over and over. Anna, on the other hand, simply stays put in the right-hand lane. Despite all of his risky lane switches, Jack survives the perils of the road, and eventually gets to work, even arriving on time. He thinks his driving maneuvers have paid off, until he sees that Anna has already parked her car and is walking into the office building.

One way or another, all drivers have experienced this. It is paradoxical, but the other lane always seems to go faster. Nevertheless, as soon as we join this "fast" lane, it turns out that in fact the lane we have just left is now the faster-moving one. Drivers who constantly change lanes jeopardize their own safety and that of others, but on average, ironically, they do not actually end up going any faster.

N. Crato, *Figuring It Out*, DOI 10.1007/978-3-642-04833-3_6,
© Springer-Verlag Berlin Heidelberg 2010

This is an intriguing problem. So much so that two statisticians decided to investigate it using a mathematical model and computer simulations. Donald A. Redelmeier, from the University of Toronto, and Robert J. Tibshirani, from Stanford University, jointly published a short article on the subject in *Nature*. Later they published a more comprehensive study in *Chance*. Their conclusions are surprising. The illusion that the other lane is moving faster is based on various objective factors. Subjective opinions are only part of the problem.

The behavior of drivers in a busy lane is never completely constant. There are always some whose speed fluctuates, some who accelerate more fiercely and some more gently, some drivers who are just slightly anxious, while others lose their cool entirely. To complicate matters, the distance between vehicles is a (non-linear) function of their speed. The greater the speed, the greater the distance between the cars has to be. The result is that, from a given volume of forward-moving traffic, a line of vehicles moves erratically, stopping and starting, even if there are no traffic lights or junctions to impede the flow of traffic. When the line starts going again, the cars do not all advance at the same time. Each vehicle only moves forward when the vehicle in front moves. When the line stops, the same thing happens, with the cars at the front stopping first and those behind only stopping later. That is why a busy line of cars has an oscillating longitudinal movement.

When two or more adjacent lanes are moving forward, with each of them oscillating in stop-and-go movements, there are moments when each car is overtaken by those in another lane and also moments when the opposite happens, i.e. when each car passes those in the other lanes. The objective of the statistical study carried out by Redelmeier and Tibshirani was to compare these moments. For this purpose they used a computer to simulate two lanes with similar movements to those experienced by real lanes full of vehicles, but they did not permit drivers to change lanes.

This simulation is not simple, and it is necessary to include the inherent randomness of each vehicle. In many traffic studies, such individual parameters are not taken into account, as only the overall

The "Stop-and-go" Phenomenon

Two adjacent lanes of traffic can be moving at the same average speed, but even so each vehicle spends more time being passed than in passing the others. This is shown by this example with two adjacent lanes moving erratically in "stop-and-go" mode. In this example no vehicles change lanes.

1st Image: the vehicles in the upper lane are passing those in the lower lane. Vehicle ❶, which had stopped, is starting to move and vehicle ❷ begins to be passed by other vehicles

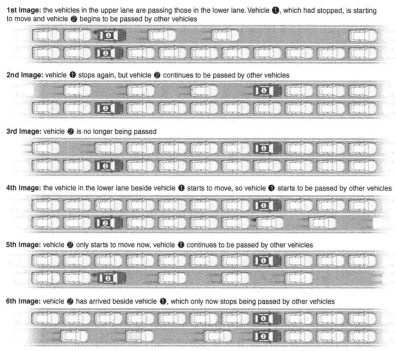

2nd Image: vehicle ❶ stops again, but vehicle ❷ continues to be passed by other vehicles

3rd Image: vehicle ❷ is no longer being passed

4th Image: the vehicle in the lower lane beside vehicle ❶ starts to move, so vehicle ❶ starts to be passed by other vehicles

5th Image: vehicle ❷ only starts to move now, vehicle ❶ continues to be passed by other vehicles

6th Image: vehicle ❷ has arrived beside vehicle ❶, which only now stops being passed by other vehicles

At the start the two vehicles were side by side, and they are still side by side at the end, but each of them was moving for less time than it was being passed

traffic flow is of interest. But it is the unpredictable behavior of individual drivers that accounts for delays and for slowing down the overall flow. That is why the study carried out by these two statisticians is so interesting.

When setting up the simulation to run on the computer, Redelmeier and Tibshirani concentrated on randomly selected individual vehicles as reference points and compared two times: the mean time taken by a vehicle to pass another, and the mean time taken to be passed. That is when the surprises started.

As long as the traffic density is low and the traffic flows smoothly, the model predicts equilibrium between the movement of the two lanes. The time during which each vehicle is passed, is, on average, more or less the same as the time that it takes to pass the vehicles in the other lane, so there is long-term equilibrium. However, when traffic density increases, the times taken to pass or be passed start to differ, as each vehicle spends more time being passed than passing others. Despite this, all the vehicles take the same average time to travel a given distance.

This is a surprising fact. How can we spend more time being passed by the vehicles in the other lane than in passing them, even though we all take the same time to travel the same distance?

Although the explanation is simple, it is still difficult to visualize. When we pass the cars in the other lane, this happens because they have stopped or are traveling more slowly. At this point their lane is more bunched up, with less distance between the vehicles, so that we can pass many vehicles quickly. Let us suppose that we pass 50 in a minute. When it is our turn to be passed, on the other hand, it is the vehicles in the other lane passing us that are traveling faster, and so are more widely spaced out. More time, maybe 2 minutes, is required for an equal number (50) of vehicles to pass us. For this reason each vehicle spends more time being passed than in passing the others, but in the end we all arrive at the same time. What's the moral of this story? The risks entailed in frequent lane changes are not usually worth the trouble!

SHOELACES AND NECKTIES

Mathematicians just love problems taken from real life that are easy to formulate. Often they turn out to be the most difficult, and therefore frequently the most interesting. This creates great enthusiasm among them for such apparently trivial questions as finding the best way to lace your shoes!

Shoe-lacing patterns have been studied by the mathematician John Halton, who considered them to be particular cases of the famous traveling salesman problem. This is a well-known and difficult mathematical problem, inspired by a real-life situation: a salesman wants to pass through a specific number of towns, visiting each one only once, but his starting and ending points are fixed. The pathway of a shoelace is equivalent to the salesman's route, with the eyes (the holes through which the lace passes) representing the towns. The shortest pathway for the shoelaces is equivalent to determining the shortest route between all the towns.

This problem was approached anew by the Australian mathematician Burkard Polster in a study published in *Nature*, one of the most prestigious scientific journals in the world. Polster systematically studied the various ways of lacing shoes.

On the surface, it would seem that there are only one or two accepted methods of lacing our shoes. However, people in different cultures tend to lace their shoes in many different ways. To take only two examples, consider the different methods normally used in the USA and in Europe. In the U.S., shoelaces are usually threaded in opposing zigzags, and when seen from above they seem to be crossed, while in

N. Crato, *Figuring It Out*, DOI 10.1007/978-3-642-04833-3_7,
© Springer-Verlag Berlin Heidelberg 2010

Europe they are typically threaded in alternating zigzags in such a way that the eyes of the shoes seem joined horizontally by the shoelaces when viewed from above. There is also the shoe-shop method in which the shoelace makes a continual zigzag from top to bottom and then returns in a diagonal line. Which of these do you think is the most efficient method?

The first curious fact is that there are actually an astronomical number of options when it comes to lacing shoes. For shoes with two rows of five eyes, for example, Polster verified that there are 51,840 different ways to thread the laces. This number rises to the millions when the number of eyes increases.

Polster restricted himself to ways of threading the laces that necessarily use all the eyes and allow them to be pulled together by applying pressure to the laces: for example, laces may not pass through three successive eyes on the same side, as this would not exert any individual pressure on the eye in the middle. Then he defined the efficiency criteria. He stated that the security of the binding should be maximized and the compression of the laces should be minimized.

Comparing the three above-mentioned systems, Polster verified that the most economical method is the American one, with the second-best system depending on the number of eyes. If there are four or more pairs of eyes, the European method is superior to the shoe-shop method. In the case of three pairs of eyes they are equal. And in the case of only one or two pairs of eyes, the problem is trivial, as all three methods are equally good. If you try to verify this, you will see that it is not difficult.

However, Polster did not restrict himself to studying just these three methods. Taking only the above-mentioned restrictions into account, he analyzed the problem and discovered that the most economical system is not, in fact, any of the three commonly used methods. Instead, he found a less well-known way to lace shoes, called the "bow-tie" method, which appears to be the most efficient of all.

As far as the criterion of maximum security of the binding was concerned, he did not find any esoteric method, which is comforting. After all, the American and shoe-shop methods are the best. When the rows

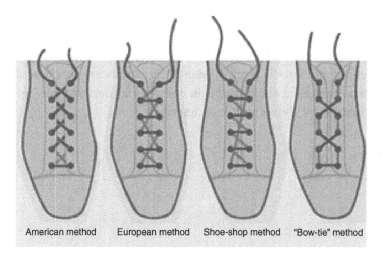

American method   European method   Shoe-shop method   "Bow-tie" method

Four different ways to lace your shoes

of eyes are farther apart, the shoe-shop method is the strongest. When the rows are close together, the American system is preferable.

Polster was probably inspired to write his mathematical work on shoelaces by an equally curious study that the computational physicists Thomas Fink and Yong Mao had published several years previously. It considered the various ways to tie a tie, a subject that gave rise to a book the pair published in 1999, called *The 85 Ways to Tie a Tie: The Science and Aesthetics of Tie Knots.*

Their study begins with a brief history of neckties and then explores the mathematical theory of knots. The two physicists try to identify every possible type of tie knot, but limiting themselves to those that can be tied in less than 10 moves. Even so, they find 85 different ways of tying a tie. The simplest of these requires only three moves. You start by placing the tie with the outward side facing your shirt, and the odd number of turns ensures that it ends up with the outward side facing outwards, as usual. This is called the "oriental knot" and is seldom used in western dress. Next is a four-move knot, which is more widely used. Things become a bit more complicated when the number of moves required to complete the knot increases. One of the more impressive eight-move knots is the Windsor, which the eponymous Duke actually never used, but is similar to the bulky knots he used to wear. It comes in handy when a larger-volume knot is desired. Many other knots are also described. However, no matter which knot is in or out of fashion, you can be sure mathematics will always be able to describe it.

NUMBER PUZZLES

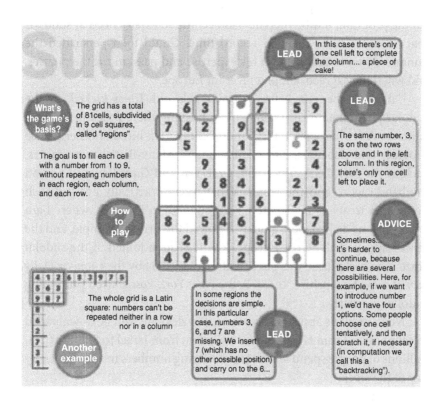

LEAD In this case there's only one cell left to complete the column... a piece of cake!

What's the game's basis? The grid has a total of 81 cells, subdivided in 9 cell squares, called "regions"

The goal is to fill each cell with a number from 1 to 9, without repeating numbers in each region, each column, and each row.

LEAD

The same number, 3, is on the two rows above and in the left column. In this region, there's only one cell left to place it.

How to play

ADVICE

Sometimes... it's harder to continue, because there are several possibilities. Here, for example, if we want to introduce number 1, we'd have four options. Some people choose one cell tentatively, and then scratch it, if necessary (in computation we call this a "backtracking").

The whole grid is a Latin square: numbers can't be repeated neither in a row nor in a column

Another example

In some regions the decisions are simple. In this particular case, numbers 3, 6, and 7 are missing. We insert 7 (which has no other possible position) and carry on to the 6... **LEAD**

The puzzle I am about to describe has an intriguing name, and one you've surely heard about. You can write it as "sudoku" or "su doku", as you prefer. It comes from the Japanese: "su" means "number" or "counting", and "doku" means "single" or "unique". In order to

N. Crato, *Figuring It Out*, DOI 10.1007/978-3-642-04833-3_8,
© Springer-Verlag Berlin Heidelberg 2010

complete this puzzle, you have to insert numbers into empty boxes. And there is only one solution.

But, contrary to what you might assume, the sudoku did not actually originate in Japan. Rather, it first appeared in the 1970s in the New York magazine *Dell Pencil Puzzles and Logic Problems*. At that time it was called "Number Place", not sudoku. In 1984, it appeared in a Japanese newspaper under the name "Suuji wa dokushin ni kagiru", which was subsequently abbreviated to "su doku", and it soon became a popular Japanese pastime. In 1997, a retired judge from New Zealand by the name of Wayne Gould became so enthusiastic about the puzzle that he began to write a computer program to study it. It took him 6 years to finish the program, but by then he had become so experienced at sudokus that he could solve them in record time. Gould then persuaded the London *Times* to make use of his expertise and passion for the puzzles, and when they agreed, sudoku fever hit Europe.

On November 12, 2004, the venerable old *Times* kicked off the sudoku phenomenon in Europe by publishing the puzzles in its daily editions. A few days later, the London *Daily Mail* hired another puzzle provider and also started to print sudokus. Many British newspapers, such as the *Sun*, *Daily Telegraph*, *Observer*, and *Guardian*, jumped on the bandwagon, and the result is that today practically every British newspaper treats its readers to a daily sudoku. In Portugal, where I am from, various daily papers quickly followed the British example, and the same happened all across Europe. The U.S. soon joined in the sudoku craze. The first puzzle was published there in 2004 by the *Conway Daily Sun* of New Hampshire. In 2005, the *New York Post* started publishing sudokus regularly.

Now, sudokus appear everywhere. Puzzles can even be downloaded to cellphones. From New Zealand to Serbia, from Israel to South Africa, millions of people spend leisure time entering numbers in sudokus every day of the week.

The sudoku is a puzzle that is typical of the 21st century. It consists of numbers, not words, it can travel the globe extremely rapidly, across language barriers. You can do sudokus on several websites (for instance on www.sudoku.com.au) and you can even compete in real time with people from all over the world (www.sudokufun.com).

Sudokus can range from the easy to the fiendishly difficult. The easiest of them can be solved by anyone in just a few minutes, while the more complex ones can take even an experienced enthusiast hours to solve. But only rarely are they so taxing that they make a fan give up in frustration.

Today, everybody is familiar with the sudoku structure. The puzzle consists of a large square divided into nine smaller squares on each side, so it has a total of 81 boxes that have to be completely filled using digits from 1 to 9 (colors or other symbols are also used in some cases). Each digit may appear only once in each row and in each column. Technically sudokus are said to be "Latin squares".

The history of Latin squares is as old as it is interesting. Apparently Latin squares were first conceived by the Swiss mathematical genius Leonhard Euler (1707-1783), in the context of problems affecting resources. Euler (pronounced "oiler") posed a problem concerning six ranks of officers and six types of regiments. He then tried to combine 36 officers within the 6 regiments in such a way that each regiment would be assigned one officer from each rank. As is normal in the case of mathematical problems, Euler formulated various conjectures with regard to these "magic squares". One of them, concerning what became known as "orthogonal squares", continued to perplex mathematicians throughout the modern age, until finally S.C, Bose, S.S. Shrikhande, and E.T. Parker solved the mystery completely in 1960.

Latin squares have also been used in statistics for the design of experiments. Ronald A. Fisher (1890–1962), for example, who is usually considered the father of modern statistics, made use of them in experiments in which three different factors were combined completely. A classic example is a study of four brands of tires fitted on four different vehicles. To prevent either the type of vehicle or the position of the tire (back or front, right or left) from interfering with the conclusions drawn from the experiment, one tire from each manufacturer is fitted to each vehicle, with each brand of tire being in a different position in each case. For instance, tire brand A is placed on the left front wheel of vehicle 1, tire brand B on the right front wheel of the same vehicle, and so on, always ensuring that two tires of the same brand are never fitted to the same vehicle and no tire brand is fitted twice to a wheel in the same position on any of the cars. Fisher then constructed a Latin square by creating a table with the rows occupied by the tires on each vehicle and the columns occupied by the four possible positions (front or back, right or left). It can be seen that the absence of repetitions favors statistical analysis, as the resistance of each tire brand is evaluated on different vehicles and in different positions. Thus the wear and tear on the tires cannot be attributed to differences between the vehicles or differences in the wheel stress, and the effects of inevitable errors are reduced.

A sudoku is a Latin square that is already partially filled with digits – the challenge for the player is to enter all the numbers to create a complete Latin square. Even before sudokus were invented, this problem had been the subject of many scientific studies. Computational scientists showed that it was a difficult problem of the "NP-complete" type. Curiously, the difficulty in solving each problem of this type depends on the number of boxes that have already been filled in. As can be easily understood, if only a few boxes are filled in at the start, the problem is easy to solve as there are many possible solutions. On the other hand, if many boxes have been filled in, there are few hypotheses left and the problem is just as easy to solve. The greatest difficulty arises in intermediate cases, a situation that is known as "phase transition". In simple Latin squares the phase transition occurs at or about the magical proportion when 42% of the boxes have been filled in. This is not exactly the case with the sudoku, as there are additional restrictions, and sudokus that have been properly designed only have one possible solution.

The great innovation of sudokus is the creation of *regions* inside each Latin square. The large square of 81 boxes is divided into nine smaller squares, each containing nine boxes. You might think that this would make the problem more complicated, but that is not true. It becomes easier and also more interesting. Why don't you try a sudoku?

TOSSING A COIN

If we toss a coin and it comes up heads, we don't find this unusual. It is just as likely that it will come up heads as tails. But if it comes up heads five times in a row, we would say that that it is unusual. And what if the order was different? Let's call heads 1 and tails 0 to make things simpler from now on. For example, if the sequence was 01001, would that also be unusual?

This problem confuses many people. Why don't you try it out? Ask a friend which of the following sequences is most likely: 11111, 10001 or 10110. You can be quite sure that your friend will say that the first sequence is the most unusual, that the second is a little more likely, and that the third is quite normal.

Well, your friend is wrong. Any one of these three events is equally likely. If you take any sequence of five tosses of a coin, the probability of it happening is $1/2 \times 1/2 \times 1/2 \times 1/2 \times 1/2$, which is about 3%. You just have to remember that it is equally likely that the coin will fall on either side each time to realize that any sequence is as probable as any other.

What happens is that we see a simple pattern in the sequence with five consecutive 1 s, and we know that this pattern is unusual. But we don't discern any pattern in the 10110 sequence, so we don't differentiate it from 10010, for instance, or from any other sequence without an apparent pattern.

If you are still not convinced, why not try this experiment. Write down a sequence of five zeros and ones. Then toss a coin five times in a row many times. You will see that it is very unlikely that your previously defined sequence will appear.

N. Crato, *Figuring It Out*, DOI 10.1007/978-3-642-04833-3_9,
© Springer-Verlag Berlin Heidelberg 2010

Mathematicians really like examples featuring coins, as they provide a simple model for discussing many complex phenomena.

We live in a world of probabilities. Very few things are absolutely certain, even things we regard as quite sure, such as court judgements. When the first forensic evidence based on comparison with DNA was introduced, some lawyers said that it did not constitute absolute proof, as scientists confirmed that there was a possibility that two persons could have certain DNA segments that coincided, though they would, of course, *de facto*, be different persons. This probability might only be about one in a hundred million, which is equivalent, when faced with confirmation by DNA analysis, to almost absolute certainty. But lawyers and judges are not used to quantifying the probability of errors in their statements and decisions, and therefore these numbers may not appear satisfactory to them. Nevertheless, one in a hundred million is very unusual indeed, so it provides a more certain proof than practically any error in a human decision.

The example using the coins can be applied to many real-life events. If you were in a European country where license plate numbers are sequential, and you saw car number AAA 111 you would be astounded. However, license plate AAA 111 (if it exists) is just as unlikely as GXF 472 or any other combination of letters and digits. Once again, it is a question of our tendency to separate what we see as a strange pattern from things that do not seem to contain any pattern whatsoever.

In our daily lives it is often difficult to consider probability from a rational standpoint. For example, we might be easily convinced that it is more likely that a traffic light will turn red as we approach it than that it will remain green. But we only think that way about it because it is annoying when a light turns red and forces us to stop, though we hardly notice the light at all when it is green and we can keep going without impediment. If we wanted to keep a careful score, we would have to define the period of time during which the event "the light turned red" could happen. In fact, if we are driving at a normal speed, the time from seeing a traffic light to reaching it can be longer than the cycle of the traffic light. In that case the light will always be red at some time as we approach it.

No, there is certainly no god of chance who randomly or consciously decides to persecute us or favour us. We only see the hand of gods when our power of reasoning plays tricks on us.

THE SWITCH

In a marvelous book written several years ago, Witold Rybczynski wondered which invention would turn out to be the tool of the millennium. After checking out various possible contenders, he chose the screw and the screwdriver. Appropriately enough, his book is called *One Good Turn: A Natural History of the Screwdriver and the Screw*.

Rybczynski's choice is arguable, just like any other. Imagine if you were asked to make such a choice. What would you select? The telephone? The airplane? The radio? Well, personally I would choose a very humble instrument indeed – the switch.

To be exact, the switch isn't really an instrument; it's more of a theoretical instrument with many varied practical applications. As an example, let us take the everyday electrical switch used to turn lights on and off. Another much more dramatic example of a switch are the points used in railroad or streetcar tracks. Older readers may well recall these. In the old days you would have seen the driver or guard get out of a carriage carrying a type of rod, walk over to the rails, and insert the rod. Then, turning it like an old-time automobile starting lever, he would manually move the points over to the desired track. This was a frequent scene in cities with streetcars. Today the railroad tracks still have points, but now they are activated by means of an remotely operated electromechanical system, although I suppose there may still be some manually operated ones left on little-used tracks.

What the train driver uses to activate these points is simply a commutator, a kind of double switch that selects the track the train is meant to take. Points are a great invention. Right at the start of the railroad

N. Crato, *Figuring It Out*, DOI 10.1007/978-3-642-04833-3_10,
© Springer-Verlag Berlin Heidelberg 2010

era, before they were introduced, railroad tracks could not be interconnected. Locomotives that ran on one track could not move to any other track. The invention of points enormously advanced the range of applications of the railroads, which in turn would be unthinkable today without these switches.

In the 19th century, as railroad lines were being developed, the switch became the basis for a means of communication that advanced in line with the railroads: the telegraph. This electrical instrument is simply the long-distance communication of signals controlled by a switch. At one end of an electric wire an operator moves a "transmitter" (a sort of crank) that connects and disconnects the electric current and thus creates impulses. Some impulses are long, some short, representing the dots and dashes of the Morse code. At the other end of the wire another operator receives these signals, which arrive in the form of audible movements of a lever. A switch at one end operates a lever at the other.

The telegraph was succeeded by the telephone. At the beginning it was only an object of curiosity, but then it became a communications system for private lines. Its inventor, Alexander Graham Bell (1847–1922), started with several telephones on his desk, each connected to a line that went to a different place. This system soon reached its limits: imagine having to have as many phones on our desks as we have friends and colleagues we wish to call! It became necessary to find a system that would be able to route the calls along the lines. In this way each person would need only one phone and one line. It was essential to create one or more telephone exchanges that could route the calls by operating the necessary switches. The first telephone exchange was inaugurated in January 1878 in New Haven, Connecticut, and was operated manually. It actually took a long time before telephone networks were automated, first using electromechanical and then electronic systems. Today's telephone system consists of a gigantic network of switches.

The symbols of our time, the computer and the internet, have resulted in the greatest concentration of switches of all, involving an extremely intricate series of connections controlled by super-rapid switches called transistors.

What makes this all so curious is that the future of computer communications depends on our ability to design ever faster switches. Today, fiber optics, which provide faster communications, are replacing electric wires. In recent years, the advances made in fiber optic technology have been vastly greater than those in computer chips. However, an emerging significant obstacle to faster communications is the relatively slow electronic control of the much faster optical signals. The future will belong to switches that work using light only, that are controlled by light, and eliminate any loss of time during the transformation of the light signal into an electric signal via electric commutation. The progress of our communications once again depends on improvements being made by that wonderful, earlier invention, the switch.

EUBULIDES, THE HEAP AND THE EURO

The euro coins have been in circulation for a few years now, so people in the eurozone should all be able to identify them. Why, then, are there still many people who get confused by them? Some people find it difficult to distinguish a two-cent coin from 5 cents, while others get the 10 and 20 cent coins mixed up, or confuse the 20 cents with the 50 cents. Is that our fault, or is it the design of the coins that is to blame for our confusion?

The designers of the euro coins decided coin sizes should increase slightly with their value. So, for example, the 20 cent coins are a little bigger than the 10 cent coins, and 50 cent pieces are also a bit bigger than the 20 cent coins. All three coins are made from the same alloy. A similar thing happens with the one and two euro coins, which have a nickel alloy, and with the one, two, and five cents, which are made from a copper alloy. It seems to be a rational system, but it doesn't yield the best results.

In my country, Portugal, when the escudo was still used, nobody got the 50 escudo coin mixed up with the 100 escudo coin, even though the 100 was smaller in diameter than the 50 escudos. In this case the logical solution of size increasing with value was completely ignored. Furthermore, these two coins were made from different alloys and in different styles, which prevented any mix-ups. This principle is used in other monetary systems too, including in US coins, for which there is no link between size and value, and different alloys are used for coins of similar values, which helps to differentiate them. The fact that in the case of the euro design the same alloys are used

N. Crato, *Figuring It Out*, DOI 10.1007/978-3-642-04833-3_11,
© Springer-Verlag Berlin Heidelberg 2010

for adjacent sizes and values of coins does not make them easy to identify.

This reminds me of the paradox of the heap, formulated more than 2300 years ago by the Greek philosopher Eubulides. Eubulides, a contemporary of Aristotle, was born in Miletus and later lived in Megara, where he was a leading member of the so-called Megarian school of philosophy. Among other celebrated paradoxes, he created the "sorites" paradox (a Greek word meaning "heap", derived from "soros", which means a hill). Eubulides started by positing that "A heap does not consist of one grain of sand, or of two, or even of one hundred grains, and we cannot create a heap by just adding one more grain of sand to a pile of grains of sand".

Everybody agreed. Then Eubulides argued "If we have a pile and we add a grain to it, we still have a pile. It seems that we will never have a heap, no matter how many times we add a grain to the original pile. . ." Now, we know this is not true: after a great number of grains is added to the pile, we get a heap. The number of grains required can be very great indeed, but we will eventually have a heap.

Eubulides then concluded "This means that at a certain point you had a pile that became a heap by adding to it one grain of sand, but this is impossible as we have agreed before."

Now, what would you reply?! That he is playing with words?! But are words not intended to express thoughts? Where is the fault in this line of thinking? In the 20th century, this paradox was considered by various philosophers, logicians and mathematicians, from Gottlob Frege (1848–1925) and Bertrand Russell (1872–1970) to some contemporary logicians. Interest in the paradox was rekindled recently by a new analysis published by the Swedish-born Oxford philosopher Timothy Williamson (*Vagueness,* Routledge, 1994).

Various solutions have been proposed for this paradox. One rational approach confirms that undefined attributes such as "heap of sand" do exist, with a zone between what is undoubtedly a heap and what is not, or not yet, a heap. But in this case it would be necessary to eliminate undefined attributes from all rigorous logical reasoning. Frege concluded that everyday language would become irretrievably paradoxical if we were to use it with its greatest intended rigor.

Another solution is to negate the validity of inductive reasoning, for example, by adopting randomized logic and taking intermediate degrees of truth in the propositions into consideration. Yet another solution would be to consider the problem as just a question of perception. Just as happened with similar heaps of sand, we get the 10 cent coins mixed up with the 20 cent coins and also 20 cents with 50 cents. And, just as happens when we compare a single grain of sand with a heap, we find it easy to tell a 10 cent and a 50 cent coin apart.

This problem is just as important in theoretical economics. In order to calculate "indifference curves" (for instance, in combinations of goods that consumers take to be equivalent), we encounter problems that are difficult to resolve in practice. If you want to buy a car, do you see any real difference in price between one that costs 15,000 dollars and one that costs 15,001 dollars? That is unlikely. But buyers beware, as the marketing professionals seem to have studied the philosophy of Eubulides. As we don't see much difference between 15,000 and 15,001, or between 15,001 and 15,002... without knowing quite how it happened, we typically end up leaving an automobile dealership driving a car with many more features and accessories than we need, having inevitably spent more money than we originally intended.

THE EARTH IS ROUND

How GPS Works

For thousands of years man navigated by the stars. But since the invention of GPS, we have replaced the Pole Star, the Southern Cross and the Sun with artificial satellites as our primary navigational guides. This may seem far less romantic, but you have to admit that the inner workings of GPS are intriguing. How does it work? Are there satellites watching us from the skies and following our every move, our every

N. Crato, *Figuring It Out*, DOI 10.1007/978-3-642-04833-3_12,
© Springer-Verlag Berlin Heidelberg 2010

position? The short answer is, not really, though the reality, while less frightening, is much more fascinating.

GPS stands for "Global Positioning System", a system first created by the U.S. Department of Defense in the 1970s and 1980s, but which is accessible to the public. The system consists of three elements: a network of satellites, terrestrial control stations, and users operating GPS receivers.

The network of satellites comprises 24 devices positioned across six different orbital planes in such a way that, no matter where you are in the world, it is always possible to receive signals from at least four of the twenty-four satellites at any given time. The terrestrial control stations monitor the satellites, always tracking their exact position in space. These stations then control the satellites so that they can, in turn, transmit their precise position. The GPS receivers are operated by users who receive signals from the satellites and calculate their position based on those signals. It is an extremely complex system.

To understand how it all works, imagine that you are on a hiking vacation in a remote area. I like to picture myself in the Alentejo region in southern Portugal, where I used to spend my summer holidays. Now, imagine you are there and have lost your way. However, you do know that you are somewhere between the villages of Ourique and Castro Verde. Then, imagine that there is a church in each of these villages and that the bells of each church can be heard for dozens of miles all around, and that both sets of church bells are rung every hour precisely, and that each has a distinctive sound, so it is easy to tell them apart.

Suddenly, you hear the church bells of Ourique. The precise time is 17.3 s after midday. This means that the sound of these bells, rung at exactly 12 o'clock, took 17.3 s to reach you. As you know that the speed of sound is about 1125 ft/s, you are in a position to calculate your distance from Ourique. Multiplying the speed by the time elapsed, you know that you are about 19,500 ft, or 6500 yards away. Seconds later, you hear the bells of Castro Verde and look at your watch: it is exactly 26.6 s after noon. Calculating the distance in the same way, you work out that you are about 29,925 ft or almost 10,000 yards from Castro Verde. So just where are you?

Now draw two circles on the map. Center the first one at Ourique with a radius of 6500 yards: this connects all the points at which you would hear the church bells of Ourique at exactly 17.3 s after noon, which means you are located somewhere on this circle. Following the same course, trace another circle centered at Castro Verde with a radius of 10,000 yards. You are also somewhere on this circle. So you have to be at a point where the two circles intersect. They intersect at two places. You are at one of those points.

You are in the middle of a field, but you notice that one of the two points of intersection marked on the map is near the freeway. Clearly, you are not there, or you would be dodging cars, and not in the middle of a vast field. Therefore, you must be at the other point. You check the map and see that you are very close to a very small village called Cabeça da Serra.

GPS works in a very similar way. The satellites emit signals in the form of electromagnetic waves (not sound waves). The GPS device receives these signals and is able to measure its distance from each of the satellites by means of wave interference. As it knows the precise positions of the satellites (recorded in the so-called *almanac*, which is continually updated), the device also knows the coordinates of your position. Your GPS device then indicates your position by pinpointing these coordinates in the map integrated into its navigation software.

If a device receives signals from two satellites, it can measure two position coordinates (latitude and longitude) after eliminating the geographical ambiguity, as once more there are two possible locations. The GPS receivers can eliminate the false hypothesis after an initialization period, during which they receive multiple signals from various satellites, just as you eliminated the location beside the freeway with the help of some additional information.

But a GPS device requires signals from four satellites. With the third signal, the receiver can measure the altitude with respect to sea level. With the fourth signal, it synchronizes its internal clock. The four signals that it receives simultaneously enable it to calculate these four coordinates. Time is like a fourth dimension.

The signals emitted by the satellites travel at the speed of light, so the measurements must be extremely precise. Atmospheric conditions have to be considered, as they modify the velocity of the waves and cause noise interference in the signals. The orbital velocity of the satellites also has to be taken into account, as this affects the frequency of the signals that are received. To increase the precision, the effects of the terrestrial gravitational field and of relative movement also have to be considered. GPS is an incredibly sophisticated and successful marriage of science and modern technology. However, its underlying principle is as simple and beautiful as two church bells pealing in the countryside.

GEAR WHEELS

Reproduction of an Antikythera Mechanism model, recently constructed by John Gleave

N. Crato, *Figuring It Out*, DOI 10.1007/978-3-642-04833-3_13,
© Springer-Verlag Berlin Heidelberg 2010

A person examining the interior of a mechanical clock cannot fail to be amazed at the number of gear wheels it contains. These gear wheels ensure that the clock's hands revolve at a certain speed by converting the oscillations of the internal energy source, usually at one-second intervals, into a one-hour cycle for the minute hand and a twelve-hour cycle for the hour hand.

To make this conversion it would be possible to construct a gear wheel with only one tooth and to connect it to the energy source, which would make such a wheel turn once per second. Then it would be possible to mesh this gear wheel with another wheel with 3,600 teeth. Each time the first wheel made a complete rotation, the second wheel would advance by one of its 3,600 teeth. When the first wheel had rotated 3,600 times, that is after 1 h, the second wheel would have completed a single rotation. So this second wheel could be connected to the minute hand.

However, a gear wheel with only one tooth would be highly unstable and fragile, just as a wheel with 3,600 teeth would simply be too massive. So in their wisdom, clock designers chose to utilize a series of gear wheels that transmit the initial movement from one to another, successively reducing their cycle of rotation. Two gear wheels are mounted on each shaft in such a way that a shaft is rotated by one wheel and simultaneously moves the other wheel, which in turn meshes with a third wheel on another shaft, and so on. The final rate of conversion of the original movement is the product of the gear ratios at each individual step. To obtain the desired rate of 3,600 : 1 there could be a series of steps, for instance, we could write the following product of fractions: $36/10 \times 50/10 \times 20/5 \times 10/5 \times 25/10 \times 100/10$, with the first number in each fraction representing the number of teeth on the first gear wheel and the second the number of teeth on the wheel with which it meshes. Obviously there are many possible combinations of ratios and so there are many possible combinations of wheels. The clockmaker's skill lies in designing an optimum series of gear wheels, neither too big nor too small, that produce the desired ratio.

In the 18th century scientists developed mathematical algorithms to solve this problem iteratively. Essentially, these algorithms factorize the numerator and the denominator of the desired ratio, i.e., they write each of the two numbers as a product of prime numbers. For example, if we want to arrive at a ratio of 28/45, we can write $(2 \times 2 \times 7)/(3 \times 3 \times 5)$ and look for various combinations. In this case, $(2 \times 7)/(3 \times 3) \times 2/5$, that is $14/9 \times 2/5$, is one of the possible solutions if you want to use four gear wheels.

The problem becomes more complicated when it is impossible or impractical to arrive at exactly the desired ratio, and only an approximation can be achieved. For instance, if the desired ratio is 997 : 1999, both numbers are prime numbers, so the simplest exact solution would be to make one gear wheel with 997 teeth and another with 1999, which does not appear to be practical as the gear wheels would be enormous. But a 1 to 2 or 10 to 20 approximation is reasonable enough in this case, as 997 divided by 1999 is 0.498..., very close to one half. Better approximations would be possible as it is known that, no matter which ratio is the final goal, it is always possible to arrive at a sufficiently precise approximation by using systems of simpler gear wheels. But solving the problem can be very laborious. After all, it was only in the 19th century that sufficiently efficient algorithms were developed to obtain systems that delivered satisfactory approximations.

Fascinatingly, up until a few decades ago it was thought that all these techniques had been mastered only recently. But in 1901, the chance discovery of fragments of a metal mechanism on the seabed close to the Greek island of Antikythera refuted this idea. When it was found, the mechanism was in very poor condition, and it was difficult to tell what it was, and therefore what its possible significance could be. There was a great deal of restorative work to do before that could be known. Finally, in 1974, Derek J. De Solla Price (1922–1983), a science historian at Yale University, solved the mystery. He concluded that it was, in fact, a truly significant discovery: it was a mechanism designed to reproduce the apparent movements of the sun and moon, including the phase changes of our satellite. It was what we today call a planetary clock or an orrery.

Planetary clocks are very rare devices, as they are costly to construct, and very complicated. Their operating system consists of gear wheels that control the movements of separate markers, usually spheres, with each marker representing a planet.

The most surprising thing about the Antikythera mechanism, however, is the precision of the ratio it uses for the lunar and solar periods. The ratio was achieved using six gear wheels, $64/38 \times 48/24 \times 127/32$, resulting in a final ratio of $254/19 = 13.36842...$ This result is correct to the third decimal place!

Until a short time ago the Antikythera mechanism only astounded mathematicians, astronomers and historians. Recently John Gleave (www.orreries.freeserve.co.uk), an English artisan, and other patient craftsmen managed to reconstruct what are believed to be replicas of the ancient mechanism. They work with the desired precision. Who could be that mysterious Greek sage who constructed that original clock 2000 years ago?

FEBRUARY 29

February 29 is a date that only comes around every 4 years. If you were born on this date, you know that your birthday only falls in leap years, that is, those years having 366 instead of 365 days, with an extra day in February. Depending on your point of view, that is either an unfortunate stroke of fate, or a reason to celebrate.

In reality, though, the situation is a bit more complicated than it may at first appear. In some cases, people born on February 29 have to wait eight whole years to celebrate their birthday. Years that are divisible by 100 are an *exception to the leap year rule:* although they are divisible by 4, and therefore comply with the general rule, they only have 365 days. That is, for example, what happened in 1900, which had 365 days.

However, if you were born on February 29, you may recall that in the year 2000 you were able to celebrate your birthday. Did somebody make a mistake when they created the 2000 calendar, or were the guiding leap year principles purposely ignored? Well, neither happened, as it turns out. As it happens, the year 2000 is an *exception to the exception:* as it is divisible by 400, it is still considered a leap year, as was the year 1600, and as 2400 will be. You may rightly wonder how such complex rules came about in the first place and whether they are necessary at all. The truth is that our current calendar, which is based on a decree first issued by Pope Gregory XIII in 1582, is now pretty universally observed, and represents the culmination of a long struggle to understand the underlying astronomical cycles and to devise a calendar that always keeps pace with the seasons.

N. Crato, *Figuring It Out*, DOI 10.1007/978-3-642-04833-3_14,
© Springer-Verlag Berlin Heidelberg 2010

The primary underlying cycle for any calendar is the solar day, which is the most obvious and most universal measurement of time, and certainly the first to be used. A second important cycle, which is in fact pre-eminent in some calendars, is the lunar cycle. A third important cycle is the solar year, which governs the annual cycle of seasons.

These cycles are not multiples of each other: there is not a whole number of days in the lunar cycle, nor a whole number of lunar cycles in a year, nor even a whole number of days in a solar year. It is not possible to have a simple and perfect calendar that will always have the same number of days in each month, will align the months with the moon, and will also have the same number of days in a year. Any calendar always favors one particular cycle, to the disadvantage of the others.

The first calendars were based on the lunar cycle. Each new moon saw the start of a new month. Early on it was realized that a purely lunar calendar was not an ideal solution for farming communities who shaped their lives according to the seasons. The ancient civilizations began to complement the lunar cycle with the seasonal cycle. However, this only made the problem worse, as a year does not contain a whole number of lunar months. In some cases, for example in the Jewish calendar, they decided to make the year variable, with some years containing 12 months and others 13 months, which means that some years had 353 days and others 385 days.

The Egyptians resolved the problem by creating a purely solar calendar. Their year contained 365 days, which was a reasonable approximation, and the days on which it started and ended had nothing to do with the phases of the moon. But the Egyptian civilization lasted a long time. As their year was about 6 h too short, over time those missing hours accumulated and became noticeable. Within a few dozen years it was clear that the official calendar was out of phase with the flood season on the Nile. After a period of 1460 years the calendar had gone through the annual seasons and returned to its starting point.

The Egyptian civilization lasted more than 4000 years, and so the astronomers in ancient Alexandria were completely aware of this error in their calendar. They suggested a simple solution: that every 4 years, an extra day be inserted into the calendar. And this is exactly what we call a leap year today.

When Julius Caesar returned from his campaign in Egypt he hadn't just been impressed by Cleopatra. He was also very struck by the Egyptians' sophisticated knowledge of astronomy. In an effort to make the Roman calendar more orderly (at that time it was utterly chaotic), Caesar summoned to Rome an Alexandrine named Sosigenes, and entrusted him with this task. The new system, which later became known as the *Julian calendar* in honor of its founder, sorted out the months, established 45 BC as year 1, and stipulated that the year would have 365 days, with an additional day every 4 years. This extra day, which was to be added between days 23 and 24 of the month of *Februarius*, was called *bissextus dies ante calendas Martii* [double sixth day before the first of March].

The Julian calendar was adopted by the Catholic Church and remained the official calendar of the church until the 16th century. The only change made was to the starting date of the calendar. Following a proposal by the 6th century Scythian monk Dionysius Exiguus, year 1 was changed to the assumed date of the birth of Christ. In Europe, many kingdoms maintained the Julian calendar, others followed Dionysius' reform.

By the end of the Middle Ages, however, it had already become clear that the calendar was not keeping pace with the seasons. The almost 11 min difference between the Julian year and the solar year had been accumulating for hundreds of years, and by the end of the 16th century the calendar was 10 days behind the solar year. The spring equinox, which should have occurred on March 21, took place on March 11. After various attempts had been made to once again reform the calendar, Pope Gregory XIII decided to take definitive action. In accordance with a proposal by the astronomer Aloysius Lilius (1510–1576), and supported by the Jesuit cosmographer Christopher Clavius (1537–1612), ten days were eliminated so that the spring equinox would again fall on March 21.

However, this time, in order to avoid the mistakes of the past, it was necessary to get rid of some of the leap years. That is why years that are multiples of 100 but not of 400 are no longer leap years. They mark the exception to the exception. Under these new rules, our calendar will only be one day out of step with the solar year in 4909. Finally, this gives us a little breathing room!

THE NONIUS SCALE

In the 16th century, sea navigation still depended on mariner's astrolabes and other relatively primitive instruments for measuring astronomical altitudes. The precision of these instruments was greatly limited by the graduated scale they used, which was normally based on a minimum unit of one degree and could be subdivided into half-degrees but not into smaller units, as the measurement marks engraved in the metal instruments had to have a certain width, and began to become indistinguishable if placed too close together.

One of the leading mathematicians of the era, Portugal's Pedro Nunes (1502–1578) thought of solving this problem by marking various scales with different units. To understand his idea, you just have to think of the subdivision of a right angle, which is equivalent to a quadrant and is sufficient for measuring astronomical altitudes. He conceived of a system consisting of 45 concentric scales marked in the quadrant, as can be seen in a replica of the instrument made by James Kynuyn on display in the Maritime Museum in Lisbon. The outer scale divided the right angle into 90 parts, i.e., in units of one degree. Inside this was a scale that divided the angle into 89 parts, i.e., into units of 90/89 of a degree. Then came another scale dividing the right angle into 88 parts, and so on until the final scale divided the angle into 46 parts. This came to be called the *nonius scale*, as Nunes was called Nonius in Latin. The nonius scale consisted of these 45 concentric scales, each with one subdivision less than the previous one.

To measure the angular altitude of the sun or another star, an observer had to hold the astrolabe or quadrant vertically and adjust the

N. Crato, *Figuring It Out*, DOI 10.1007/978-3-642-04833-3_15,
© Springer-Verlag Berlin Heidelberg 2010

A modern reproduction of the Kynyun instrument, the only extant instrument from Nunes times with the original nonius scales

alidade (the device that allows one to sight a distant object and fix that line of sight) so that it was aligned with the position of the star. The angle was then measured, but this measurement was not restricted to just one scale, as in traditional instruments. The alidade passed through all the concentric scales and the observer would select the mark that coincided best with the position of the alidade. Let us suppose, like Pedro Nunes in his masterpiece *De crepusculis*, that the mark that best coincided with the position of the alidade was on the fourth scale from the top (the one that divides the right angle into 87 parts). And let us imagine that it coincided with mark number 30 on this scale. In this case the measured angle would be 30/87 of the quadrant. A fraction of 30/87 is approximately equivalent to 31° 2′ 4″, a value obtained with greater precision than in any of the individual scales used.

The nonius began to be studied all over Europe. The greatest astronomer of the time, the Dane Tycho Brahe (1546–1601), constructed various quadrants using scales like the nonius scale: "I used the subtle process presented by Nunes", said Tycho, "and made it more exact, increasing the number of subdivisions and calculating tables".[1] The astronomer eventually admitted that the instrument did not provide the precision that he wanted. So other practical solutions to improve the system continued to be sought.

A German mathematician named Jacob Kurz, an influential figure in central Europe and in the Vatican, proposed a modification of the nonius scale, which would introduce simpler, more graduated scales. Then, the Jesuit mathematician Christopher Clavius (1538–1616) took up Kurz's idea and further refined it by limiting the nonius scale to two scales. In his *Geometria practica* published in Rome in 1604, he suggested that one scale should consist of units of one degree while the other should be marked in units of one and one-sixtieth of a degree, i.e. of one degree and one minute ($1° + 1°/60 = 1°1′$).

Using this process we can directly obtain measurements of degrees of minutes by comparing the two scales. The distance between the first mark on the primary scale and the first mark on the secondary scale represents one minute of one degree. The distance between the second marks on the two scales is equivalent to two minutes of one degree, and so on. According to Clavius, we start by using the primary scale to measure the angle. The result is a certain number of degrees as well as a small remainder. To measure this residual fraction of one degree, Clavius suggested using a pair of compasses to compare it with the differences between the marks on the two scales. If the residual amount is equal to the difference between the tenth marks on the two scales, for example, this would mean that that amount is equivalent to ten minutes of one degree.

---

[1] Letter to Christopher Rothman, January 20, 1587, as quoted by A. Estácio dos Reis, O nónio de Pedro Nunes, *Oceanos* 1988, p. 72.

This process was ingenious in its logic, but was still not very practical. Another 30 years went by before someone else came up with a quick method for transferring measurements from one scale to the other. This was the innovation made by a French mathematician called Pierre Vernier (1584–1638) in 1631, and it involved adding a moveable secondary scale to the instrument, allowing the direct measurement of the residual angle imagined by Clavius.

Vernier's invention was very successful and was quickly adopted by instrument-makers all over Europe. It was tailored for use in sextants, setting circles for telescopes, callipers, gauges, and other instruments. In fact, it is this version of the nonius scale that we use today.

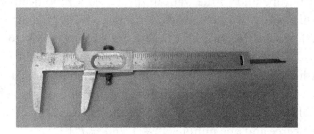

A modern calliper with a moving nonius scale built according to Vernier's idea

PEDRO NUNES' MAP

When you fly from, say, my home, Lisbon to New York, you usually reach the U.S. coastline at least an hour before your plane lands. During this hour you can typically look down and see the indented outline of the Massachusetts coast and the island of Martha's Vineyard, Nantucket, and other landmarks. Then you will soon see Long Island and all its beaches, including those of the Hamptons, as you travel northeast to southwest towards New York City. If you were to look at a map of this route, it would seem that the plane had taken a long way round and that, instead of taking the shortest route across the Atlantic, it had reached the coast of the New World farther to the north, where the Portuguese first landed, and then followed the coast.

However, if you look at a globe, you will see that the plane did in fact take the shortest route, which is the arc of a great circle, or in other words of a circle that connects the points of departure and arrival and that is centered at the center of the Earth. If you stretched an elastic band around the Earth, making it touch Lisbon and New York, you would see that the shortest route between the two cities does in fact touch the Massachusetts coast. That means that to take the shortest path between these two places, which share very similar latitudes, (Lisbon is at 39°, New York at 41°) the airplane begins by flying to the west and a little north, and ends up flying west and a little south.

All this may seem simple and obvious, but it took navigators many years to comprehend it. The first person to understand it in all of its implications was our friend Pedro Nunes (1502–1579), creator of the nonius, and the Portuguese Royal Cosmographer.

N. Crato, *Figuring It Out*, DOI 10.1007/978-3-642-04833-3_16,   65
© Springer-Verlag Berlin Heidelberg 2010

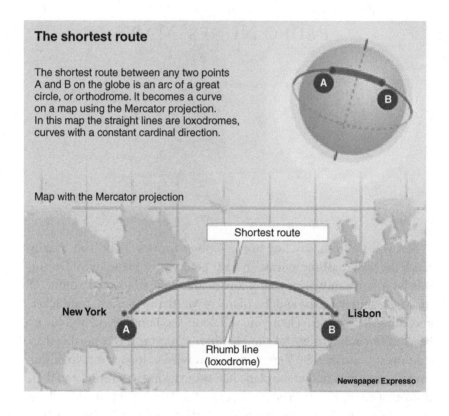

The shortest route

The shortest route between any two points
A and B on the globe is an arc of a great
circle, or orthodrome. It becomes a curve
on a map using the Mercator projection.
In this map the straight lines are loxodromes,
curves with a constant cardinal direction.

Map with the Mercator projection

Shortest route

New York

Lisbon

Rhumb line
(loxodrome)

Newspaper Expresso

In tackling the problem Nunes first turned it on its head. He asked: should you always take the same direction in order to travel most efficiently between two points? This problem had first been posed by Martim Afonso de Sousa, the founder of Portugal's first colonies in Brazil. Wishing to travel from the Rio de la Plata in Brazil to Lisbon, he ascertained that simply steering east was insufficient; that he would also have to steer north. And he realized that it was not a simple matter to chart the exact direction in which to travel.

The brilliance of Pedro Nunes is hard to overestimate. In 1537 he was able to clearly distinguish two different trajectories for a ship on the high seas. One would be a minimum-distance trajectory between two points, which corresponds to an arc of a great circle: this is called an *orthodrome*. The other would be the trajectory followed by a ship that always maintained the same orientation with respect to the cardinal points: this is called the rhumb line, later known as a *loxodrome*. These two trajectories are only identical when the ship is traveling along the Equator or along a meridian. In every other case they are different.

When he discovered the rhumb line, Pedro Nunes also proved that, on a hypothetical planet completely covered by water, a ship that always followed the same cardinal direction would not return to its starting point, as was thought at that time. Instead, it would travel in an endless spiral, getting ever closer to one of the poles by making an infinite number of turns around it. This curve, which is "neither a circle nor a straight line" as Nunes said, came to be known as a loxodromic spiral.

Undoubtedly the most spectacular illustrations of loxodromic spirals have been penned by Maurits Cornelis Escher (1898–1972). In 1958, Escher created some drawings of spheres with spirals, apparently without any knowledge of their profound historical and geometrical significance. The title of the image reproduced here is *Sphere Spirals,* and it is one of his most beautiful creations. It is a woodcut in four blocks (one for each color) with a diameter of just over 12 inches. We can imagine a ship's captain who decides to constantly steer a course of about a 60-degree angle to the north-south axis. The spirals show the path that would be taken by such a ship. And they show that, even if the ship started from different points, its trajectories would converge. The bands are broader at the Equator and become narrower as they approach the poles.

As well as showing navigators the path that they will take if they steer a course at a constant angle to a cardinal direction, the loxodromes of Pedro Nunes had a major influence on how maps began to be drawn, and greatly contributed to the vision of the continents that we have today. Nunes recognized that with his discovery, the old sea charts would have to be replaced, and in 1566 he clearly explained the precepts that should be followed in drawing the new navigational maps. From 1569 onwards, Gerardus Mercator (1512–1594) would usher in a revolution in cartography inspired and guided by the discoveries of Pedro Nunes. Mercator was born in the town of Rupelmonde in

Sphere spirals by E.C. Escher

Flanders (present-day Belgium). He was baptized as Gheert Cremer, but his name was later Latinized to Mercator (Flemish "cremer" = English "merchant" = Latin "mercator"). He first studied in Holland and later in Leuven, in Flanders, where he remained and dedicated himself to constructing globes and maps.

Mercator was, of course, aware of the work of Pedro Nunes, as the knowledge had spread throughout Europe and had been discussed by their mutual friend John Dee (1527–1608), Queen Elizabeth's astronomer and astrologer. Mercator resolved to design a map that would be of immediate use to navigators. He decided to form a grid in which the lines of latitude would all be parallel to the Equator and perpendicular to the meridians, which would run parallel to each other. He also decided that the rhumb lines should appear as straight lines, which had initially been proposed by Nunes. To achieve this, Mercator progressively increased the distances between the parallels as the lines of latitude approached the poles. In this he was again following Nunes, who had spoken of the necessity of using "increased latitudes" for drawing a map in which the loxodromes were straight lines. This is how the "Mercator projection" was created, which even today is still the best known and most utilized navigational mapping method. Its great strength derives from being a *conformal map*, as it preserves the direction between any two points on the globe. It is ideal for planning and plotting rhumb courses.

However, as with any map projection system, Mercator's method inevitably leads to distortions. If you carefully peel an orange to obtain a whole skin and then try to flatten it on a table, it will break into pieces and become bent. In the same way, cartographers have to take liberties with the geometry of the globe in order to reproduce a spherical surface on a plane. These deformations are of minor importance when only a limited surface area is considered. But the world had increased in size with the voyages undertaken in the Age of Discovery, and these distortions soon became significant.

As Mercator's maps are so widespread, their inevitable distortions have molded our sense of geography. On these maps, Greenland looks enormous, even bigger than South America, when in fact the land mass of South America is nine times larger than Greenland's. Mercator's map also continues to deceive modern travelers, who are amazed at the routes flown by airplanes today.

LIGHTHOUSE GEOMETRY

Strolling along the coast of the sea on a late summer evening, we can sometimes discern the flash of a lighthouse in the distance, blinking intermittently, as if trying to send us a signal. And indeed it is! The lighthouse is telling us its name. It is sending us a message that is known technically as its *light characteristic*.

Some lighthouses flash rapidly, while others send out prolonged signals. Some have red lights, others white lights, and some alternate colors. The pilots of ships approaching the coast are trained to read these signals, and so to identify the lighthouses that they encounter.

The code lighthouses utilize is simple, consisting mainly of three elements. The first is the way the light is sent. If the lighthouse sends out short signals, they are called flashes, a term abbreviated to "Fl". If the lighthouse emits an almost continuous light interrupted by short periods of darkness, this is known as "occulting", abbreviated to "Oc". If the light is blocked and released for equal periods, this is called "isophase", usually abbreviated to "Iso".

The second element of the code is the color of the light, which is usually designated by its initial (R = red, W = white, G = green, etc.). The third and final component of the code is what's called the period, that is, the duration in seconds of a light cycle.

These characteristics are so distinctive that they can clearly identify the lighthouse itself. For example, if you are close to the estuary of the river Tagus near Lisbon, in Portugal, you will see the light emitted by two different lighthouses, one at Bugio, and another at São Julião da Barra. The former is marked on maps as "Fl G 5 s", which means that

N. Crato, *Figuring It Out*, DOI 10.1007/978-3-642-04833-3_17,

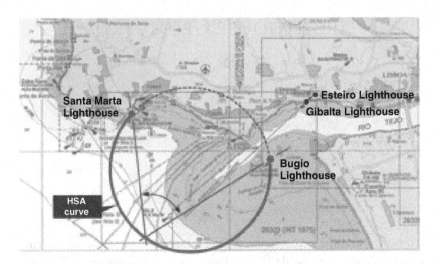

The pilot of a ship cannot calculate its position merely by observing the angle made by two lighthouses. The set of points that comply with this angle is an arc of position (known technically as a "horizontal sextant angle curve" or HSA curve). The pilot requires additional information, such as a third lighthouse, to resolve the problem

its light flashes (Fl) colored green (G) at intervals of 5 seconds (5 s). The latter is marked as "Oc WR 5 s", which signifies that it emits an almost continuous light that is blocked briefly (Oc), displaying a white color at one position (W) and red in the other (R), at 5-second intervals (5 s).

Lighthouses provide invaluable information enabling a ship to locate its position. Even today, in the age of GPS and automatic navigation, they still serve a vital function, making it easier to navigate visually along a coastline, which provides ship captains with an additional security check and permits them to plot the position of the boat. Lighthouses can also be used as guides for setting a course. At the entrance to the Tagus estuary, for example, captains can align their ships by the lighthouses at Gibalta (close to Caxias) and Esteiro (in the grove of trees around the National Stadium), thus ensuring that they are charting a safe course for their ships where the river is deep enough.

At times captains also take the angular height at which lighthouses appear to them above the horizon, and knowing the actual height of the light beam, they can apply some elementary rules of trigonometry to estimate their proximity to the coast, or they can measure the angles between various lighthouses to determine their ship's position. Lighthouses are useful to sailors in a thousand and one different ways.

ASTEROIDS AND LEAST SQUARES

Carl Friedrich Gauss, the mathematician who created the "least squares" method, which is commonly used in science

Eugene Wigner (1902–1995), who was awarded the 1963 Nobel Prize for Physics, wrote an article in 1960 that has since become a classic, titled: "The Unreasonable Effectiveness of Mathematics in the Natural Sciences". In Wigner's words: "There is a story about two friends, who were classmates in high school, talking about their jobs. One of them became a statistician and was working on population trends. He showed an scientific article to his former classmate. The article included, not unusually, the Gaussian distribution. The statistician explained to his former classmate the meaning of the symbols for actual population, for average population, and so on. His classmate was a bit incredulous and was not quite sure whether the statistician was pulling his leg. 'How can you know that?' was his query. 'And what is this symbol here?' 'Oh', said the statistician, 'this is pi.' 'What is that?' 'The ratio of the circumference of the circle to its diameter.' 'Well, now you are pushing your joke too far', said the classmate, 'surely the population has nothing to do with the circumference of the circle'".[1]

Experts may find it quite natural that π (pi) appears in population studies, as it is regularly employed in normal or Gaussian distribution, which is, in turn, frequently used in statistics. But even mathematicians would have difficulty offering a simple and convincing explanation for why this symbol has become indispensable in so many areas of mathematics and the sciences. And the relationship between π and population studies is even more surprising.

People have often been surprised by the extraordinary effectiveness of mathematics in describing, comprehending and forecasting natural phenomena.

In the early years of the 19th century the effectiveness of mathematics was demonstrated in spectacular fashion. At that time astronomers were searching for a previously unknown planet they suspected might be located between Mars and Jupiter. In fact they had detected a regularity in the positions of the planets, called the Titius-Bode law, and verified that there was a large void between Mars and Jupiter. They thought that this void might be filled by an as yet undiscovered planet.

Astronomers cooperated internationally in a group they jokingly termed the "Celestial Police". This was the first time that a scientific investigation had been so organized at the international level. Each astronomer assumed responsibility for one sector of the sky. The zone of the zodiac in which the sun, the moon and the planets describe their apparent movements was divided into segments so that nothing was left out. The sky was observed night after night by astronomers from different countries, all looking for a luminous dot moving slowly against the backdrop of the stars.

---

[1] See, e.g., http://pascal.iseg.utl.pt/~ncrato/Math/Wigner.html

The honor of discovery was claimed by Giuseppe Piazzi (1746–1826), the director of the observatory in Palermo in Sicily, who detected an unknown object, apparently at the orbit of the desperately sought "fifth planet". His discovery was made on January 1st, 1801, the first day of the new century.

Piazzi was cautious and would only admit that it might be a comet. But the members of the Celestial Police had no doubts. They were certain it was the missing planet between Mars and Jupiter. The problem was that the object soon disappeared in the darkness, and then reappeared too close to the sun to be accurately observed at that time. In the early fall of 1801, when the object ought to have reappeared in the early-morning sky, they could not find it. Heinrich Olbers (1758–1840) and various other astronomers did their calculations but still could not locate the celestial body. At this point a German mathematician entered the picture. Although only 24 years old at the time, Carl Friedrich Gauss (1777–1855) was already considered one of the greatest geniuses in the history of mathematics. The young Gauss exulted in the possibility of putting his theory into practice, as he had spent several years studying the problems involved in calculating orbits "without any theoretical assumption, from observations not embracing a long period of time",[2] as he wrote.

Gauss calculated the orbit on the basis of a new method of combining observations and using these combinations to estimate the parameters of a function, in this case an orbit. This procedure later became known as the method of least squares, and it solved a problem that the best minds in Europe had been debating for decades.

The calculations made by Gauss provided estimates for the orbit of the object that were so precise that Franz von Zach (1754–1832) succeeded in rediscovering it on December 31st, 1801, almost exactly 1 year after Piazzi had first sighted it. He discovered it at half a degree of angular distance from the position predicted by Gauss. The following night Olbers saw it too. The scientific community was euphoric. It seemed that the solar system was complete again!

---

[2] See William Sheehan, *Worlds in the Sky: Planetary Discovery from Earliest Times Through Voyager and Magellan*, Tucson and London: The University of Arizona Press, 1992, p. 105

At that time it was thought that the object was a planet similar to the others in our solar system, and it was given the name of Ceres. But in 1802, when its diameter was estimated, it was measured at just over 160 miles, not large enough to qualify as a planet. (Now, of course, we know that the diameter of Ceres is about 600 miles, almost a third that of the moon). But surprises continued apace in the following years, which saw the discovery in the same area of Pallas (1802), Juno (1804) and Vesta (1807), which is the most brilliant of all these small celestial bodies and can even be observed with the naked eye. In the ensuing years many thousands of asteroids have been discovered, and about 40,000 of them have been cataloged to date.

How is it possible that π, used in Gaussian distribution to reduce diverse observations to a simple equation, could have helped to discover Ceres and many other celestial objects? The unreasonably effective science of mathematics is the mysterious grammar of modern scientific knowledge.

The Useful Man and the Genius

If you took the time during the 2009 International Year of Astronomy to observe the sun, you would likely have noticed that our star was then without its famous dark spots. This lasted for some time. The sun was spotless on 266 of the 366 days in 2008, and all the way up to October of 2009 there were still virtually no spots to be observed. It is not uncommon for the sun to appear spotless for a brief time, but it is unusual to see it without sunspots for such a prolonged period. That had not occurred since 1913.

Sunspots are gigantic magnetic storms that spew forth material, cause sudden changes in the magnetic field, and emit intense radiation in the ultraviolet range. They are dark, but solar activity is most intense at their edges, so the total radiation emitted by the sun increases. The fact that we were registering a longer than usual period of low activity meant that less solar radiation was reaching us, which in turn undoubtedly had an impact on our climate. Maybe we will soon be able to measure it.

Great caution has to be taken when observing sunspots. It is extremely dangerous to use binoculars or a telescope to observe the sun directly, as this can result in immediate blindness. Professional filters must be used to ensure safety. The simplest and most practical way to observe sunspots without endangering vision is to project an image of the sun on a white surface (for example, by using an inverted ocular) and to observe that image. If there are any large spots, it is easy to detect them using this method.

N. Crato, *Figuring It Out*, DOI 10.1007/978-3-642-04833-3_19,
© Springer-Verlag Berlin Heidelberg 2010

The discovery of sunspots at the beginning of the 17th century by the German theologian Fabricius, the Italian physicist Galileo, the Jesuit astronomer Scheiner, and others was for a short time a major sensation. Decades after the excitement had abated, sunspots were relegated to the category of a curiosity, with little continuing significance for astronomy. But in 1844, Heinrich Schwabe, a German pharmacist and amateur astronomer, noticed a regular pattern: he discovered that sunspots increased and decreased in cycles of about 10 years. Then Johann Rudolf

Johann Rudolf Wolf (7 July 1816 – 6 December 1893)

Wolf, the director of the Observatory in Bern, Switzerland, took note of Schwabe's observations and decided to study the phenomenon more closely. He spent the rest of his life counting sunspots, which he did day after day whenever the meteorological conditions permitted. He collected various observations made in the past, and invented a method for quantifying sunspots, beginning to document a long sequence that is still being quantified today, and that reveals the oscillating pattern of solar activity. In 1852 he measured solar periodicity using an elementary statistical method and found the period to be 11.11 years.

An entire life devoted to counting sunspots may seem like a trivial calling, but as Wolf said later, "I have always consoled myself that he such as I who is not a genius, can still achieve much that is useful when he does his work right and chooses his work to suit his talents".[1]

Some 62 years later, at the start of the 20th century, a young German physicist studied the observations made by this useful man and devised another method for estimating sunspot periodicity, one based on a sophisticated mathematical tool called the Fourier transform. It was the first practical application of what is today called spectral power analysis.

That young physicist was Albert Einstein, and he also calculated the sunspot period as 11.11 years. Sunspot activity will resume and will again fade. We know that thanks to Wolf, a remarkably humble man, and to Einstein, a remarkable genius.

[1] A.J. Izenman (1983). J.R. Wolf and H.A. Wolfer. A historical note on the Zurich sunspot relative numbers, *Journal of the Royal Statistical Society* A **146**, 3, 311–318.

SECRET AFFAIRS

ALICE AND BOB[1]

"How am I going to tell Bob I love him?"   "I can't wait to read Alice's letter."

Alice and Bob live apart from each other and can only communicate by "snail" mail. But they know that the mailman reads all their letters. Alice has a message for Bob and doesn't want the mailman to read it. What can she do? She has already thought of having the message delivered in a padlocked box. But how can she get the key to Bob? She can't send it inside the box, because then Bob couldn't open the box.

After giving the problem a lot of thought, she has a brainstorm. She does send him the padlocked box. She knows Bob well enough to know he is intelligent and will eventually figure out her brilliant idea. With the mail going back and forth a few times, but without ever exchanging keys, the message arrives at its destination, where Bob is able to open the

[1] This article, together with the next two articles, was awarded the first prize in the 2003 *Raising Public Awareness of Mathematics* competition organized by the European Mathematical Society. They are reproduced here with some minor changes.

box and read the message. How do you think they solved the problem? If you like logical challenges, take a break now and think about it.

It is quite simple... Bob receives the box. When at last he understands Alice's stratagem, he locks the box using a second padlock to which he has the key. He sends the box, now locked by two padlocks, back to Alice by mail. She then removes her padlock using her key, and sends the box off again by mail. When Bob receives it again, he only has to open his padlock with his key and read the message. The mailman is left guessing.

I have just described an old brainteaser and one of its solutions. In 1976 it inspired three young Americans, Whitefield Diffie, Martin Hellman and Ralph Merkle, to design a cryptographic system in which the secret to be communicated is secured by two keys that the participants do not exchange.

THE ORIGIN OF PUBLIC KEYS

The process invented by Diffie, Hellman and Merkle marks the start of cryptography using public keys that work in conjunction with secret keys that are never exchanged. It is based on modular arithmetic, which essentially consists of working with the remainders of entire division by a specified number called the modulus. The best example is provided by a clock. If the clock shows 10 o'clock at a given moment, what time does it show five hours later? Obviously the answer is 3 o'clock, which is equivalent to the remainder of the whole division of $10 + 5 = 15$ by 12. In mathematics this is written $10 + 5 \equiv 3 \pmod{12}$, as $15 \equiv 3$ in the modulus of congruence 12 in normal analog clocks. We use this notation to describe the process adopted by Alice and Bob, using an example provided by Simon Singh. Our two friends manage to agree a common cryptographic key without ever personally exchanging it and without anybody else being able to discover it.

Although Alice and Bob are fictitious persons, these names have become standard terms used by specialists in cryptography. It is more fun to use these names than to always talk about the sender and the recipient, or only about A and B. They are usually joined by a third person (in our story that was the mailman), who is normally called Eve and plays the role of the avid listener or *eavesdropper.*

ALICE	BOB
Alice and Bob agree on the numbers 7 and 11, so they will calculate the result of 7^x (mod 11). (They do not bother to keep this information secret).	
Alice selects **3** as her secret number.	Bob chooses **6** as his secret number
Alice calculates $7^3 = 343 \equiv 2$ (mod 11).	Bob calculates $7^6 = 117649 \equiv 4$ (mod 11).
Alice sends the result, **2**, to Bob.	Bob sends the result, **4**, to Alice
(This is usually a crucial moment that the participants try to keep secret. Nevertheless, this is not a consideration in this example. Even if this exchange of information became public knowledge, nobody would be able to find out the secret key.)	
Alice takes Bob's result, 4, and her secret number, 3, and calculates $4^3 = 64 \equiv 9$ (mod 11).	Bob takes Alice's result, 2, and his secret number, 6, and calculates $2^6 = 64 \equiv 9$ (mod 11).
Alice and Bob end up with the same number, **9**, without either having informed the other of their personal secret numbers.	

Until this system was discovered by Diffie, Hellman and Merkle, the communication of encrypted messages required the code key to be exchanged. It was necessary that Alice and Bob met previously and agreed on a key that only they knew. Only this permitted them to subsequently exchange messages at a distance without Eve, always on the lookout, getting to know them. This is how secret messages have been sent and received from the time of Caesar right up to the modern era; this is how spies, governments, generals, conspirators and even lovers operated. The key might be simple, but it was always necessary that Alice and Bob agreed on the entire system for coding and decoding messages from the outset. Usually though, it was possible to compress all the information into a single number, quite possibly a very large number, so it could be said that the key consisted only of such a number.

So the idea of Diffie, Hellman and Merkle was revolutionary. In accordance with the method they laid out, Alice and Bob began by agreeing on two numbers. These numbers could even be public knowledge, since even if Eve managed to get hold of them, she would not be in a position to deduce the key. Then Alice and Bob each choose another number that they keep entirely secret. After performing some calculations, they both come to the same result: a number that nobody else knows and that will become the key for encrypting their messages. The process they invent is relatively simple, but very ingenious, and is described briefly in the text box. It all happens as described in the story of the two padlocks. The keys are not exchanged, but in the end both Alice and Bob are able to open the box. The mailman (Eve) cannot.

In a highly entertaining and enlightening work entitled *The Code Book* (Anchor, 2000) Simon Singh writes that it all happens as if Alice and Bob had wanted to invent a secret paint without anyone knowing all their ingredients. They start by choosing a certain color and then each of them puts a quart of this color in a can. Back home, Alice adds a quart of a secret color that she doesn't tell anyone about, not even her partner. Bob does the same. Then they exchange the two cans with the mixtures, not caring whether Eve is watching or not. Each of them takes home the can with the two quarts of color resulting from the addition of the secret color to the other color. Back home again, Alice adds a quart of her secret color to the can that Bob gave her. So she now has three quarts of paint: one third is their original agreed color, one third is Bob's secret color and the remaining third is her own secret color. When Bob gets home he does the same with the can that Alice gave him. The result is identical to the color Alice obtained, as the ingredients of the paint are the same. Neither told the other about their secret choice, but they ended up with the same final color of paint, without anybody being able to discover their secret, not even Eve, who was always sniffing around and who had seen them exchange the cans.

There aren't any colors in encryption systems, but there are numbers, as well as the ingenious application of a branch of mathematics known as number theory. Without these advances in cryptography, online commerce and communications would not be as secure as they are today.

INVIOLATE CYBERSECRETS

Are you apprehensive about sending your credit card number to a web-site? Is there a CD or book you decided not to buy because the seller required you to use Visa or MasterCard on the Internet? Well, you are certainly not the only person who doesn't trust the web. Many people around the world still do not take advantage of the efficiencies of e-commerce because they do not have confidence in the security of online transactions. But you might be surprised to find out that sharing confidential information through the web is actually one of the most secure transactions ever devised. If you take some elementary precautions, such as having nothing to do with sellers who are not known to you and not sending confidential information by normal (i.e. non-encrypted) email, the world of e-commerce is at your fingertips.

The Internet opens up possibilities that people did not even dream of just a few years ago. It has become a giant public library offering people everywhere rapid and secure access to international commerce. Do you want to buy that technical manual that you just can't find in the bookshops (possibly because for the life of you you can't remember the exact wording of its title)? Do you want to acquire that Bob Dylan CD that you have spent so much time looking for? Are you a collector searching for a 19th century compass? The internet could help to gratify all these wishes, as it provides access to various international chains of second-hand bookshops. One of them might just be offering that long out-of-print autobiography of Max Planck that you have been searching

N. Crato, *Figuring It Out*, DOI 10.1007/978-3-642-04833-3_21,

for. Who knows, you might even find a second-hand bookshop in New Zealand selling it at a bargain basement price on the Internet, as I did.

But is it safe to send a credit card number via all those *bits and bytes* that end up God knows where? How can Amazon and all those other online stores guarantee that your data will not end up in less scrupulous hands tapping the keyboard of some PC in Cochinchina? They may well claim that the data is encrypted, but if my message is encoded using a key that they send to my computer, couldn't some scoundrel find out the details of this key?

This question makes a lot of sense. For thousands of years secrets that were communicated were encrypted using a system based on a symmetric key, which allows messages to be coded and decoded. The persons who want to share the secret agree on the key. For example, they might agree that A is to be written as B, and B as C, and so on. So if they wanted to write GOOD DAY, the coded version would be HPPE EBZ. The key that allows this message to be encoded can be reversed to decode it. The security of the message depends on keeping the key secret.

Encrypted communication on the Internet is based on an innovative method, the asymmetric key. It is a really revolutionary step in cryptography, maybe even the most significant advance since coded messages first appeared. The method, which is incorporated in browsers, in some email systems and, of course, in interbank communications, is based on a suggestion made by Ronald Rivest, Adi Shamir and Leonard Adleman, three scientists from the Massachusetts Institute of Technology, who proposed an encryption method in 1977 that became known by their initials: RSA.

Using this system, the recipient of the message, for instance the internet vendor, creates a key consisting of two large numbers (N, e). He sends these two numbers to his client's computer without paying any attention to their security. If he wants, he can even publish them in a newspaper. Then the client's computer rewrites the message it wants to send as a numerical sequence (normally in accordance with the ASCII code), obtaining a third number (M), and then applies a simple equation: it raises M to the power of e, divides the result by N and calculates the remainder, obtaining the number C, which it sends back on the internet.

The astounding thing is that this number C, which is the encrypted message and contains data such as the credit card number, can be viewed by anybody, because, even if they possess the public code (the numbers N, e), the message cannot be decrypted. This is due to the fact that the mathematical function that transformed the number M into the number C is not one-to-one: it transforms the original number into a perfectly determined number, but other numbers could also have produced the same result, so no Internet huckster gets to know our credit card number.

So how can the recipient of the message decrypt it? Well, the recipient, who sent the public key, knows how it was created: he selected N as the product of two prime numbers (numbers that are only divisible by 1 and by themselves), let us call them p and q, and he did not reveal them to anyone. Knowing them, he calculates another number, d, in such a way that $(ed - 1)$ is divisible by $(p - 1)(q - 1)$. Then he raises the encrypted number C to the power of d, divides it by N and calculates the remainder of the entire division. This remainder is the original message, M. A miracle? Not at all. It is simply the ingenious application of a result from number theory known as Euler's theorem.

What makes this method practically inviolate is that it takes an extraordinarily long time to factorize a number that is the product of its prime numbers if these numbers are large. It is easy to obtain the product of two numbers. But even if you know that a given number is the product of the two primes p and q, finding these is anything but simple: and without knowing them, the message cannot be decrypted.

It is sufficient that the number N is large and that its two factors p and q have been carefully selected to ensure that the time required by a computer to factorize them is extremely lengthy, so lengthy that it is not a practical possibility for any cyberpirate to attempt this feat. For example, if each of the two factors has 100 digits, and a computer has the capacity to make a trillion attempts per second, even so the estimated period of time since the beginning of the universe would not be enough to ensure that such a computer would succeed in finding out the prime factors of this number.

As always, each advance in cryptographic techniques is followed by an advance in cryptanalytical techniques. The RSA method has been subjected to various attacks by mathematicians who have attempted to devise an algorithm to decrypt the private key d, in some cases by means of the factorization of the public key N to its prime factors. The successful outcomes that have been achieved have imposed restrictions on the choice of the system components and have obliged the experts to use various stratagems to optimize the security of the system, namely large numbers for the RSA keys. Until now mathematicians have not succeeded in developing any form of decryption whatsoever that can break the RSA code. As it turns out, e-commerce is still much safer than hiding money under your mattress.

Quantum Cryptography

The integrity of bank transactions, e-commerce and military signals is ensured by the utilization of very secure cryptographic systems. Very secure, however, does not mean absolutely secure. The security of the most reliable modern cryptographic systems, the RSA system, hinges on the difficulty of determining the prime factors of very large numbers.

The algorithms that have been devised up to now have not succeeded in performing this operation within a reasonable time, even using the most powerful computers currently in existence. But if some mathematician discovers an effective procedure for performing this factorization, or if a new generation of computers is introduced (the "quantum computers" on which many scientists are currently at work), then the world of communications as we know it today could collapse. If one of these revolutionary innovations was suddenly available, e-commerce would cease to be secure, the military would have to review all its communications systems, and banks would have to take a step back in time and conduct their transactions at a snail's pace. It could end the information society as we have come to know it.

What we all need is a new form of cryptography that is 100% secure. By the time that RSA begins to be vulnerable, mathematicians, physicists and computer scientists hope to have put in place a new system that will take its place, and be impenetrable. This will be so because the new system's security will employ the most basic laws of matter, the uncertainty principles that are at the heart of quantum physics. The absolute impossibility of predicting the behavior of elementary particles will ironically guarantee the privacy of messages under the new system.

N. Crato, *Figuring It Out*, DOI 10.1007/978-3-642-04833-3_22,
© Springer-Verlag Berlin Heidelberg 2010

This idea has already been germinating for some time in the minds of researchers. Charles Bennett is one such scientist. Together with a colleague named Gilles Brassard at the IBM Center in New York, in the 1980s he finally succeeded in conceiving a quantum-based cryptography system. But for years afterwards, in fact until the beginning of the 21st century, this all remained in the realm of ideas and experimentation. Recently, though, so much technical progress has been made that it has become possible to put prototypes of truly secure cryptographic systems into practice.

One of these cryptographic systems, which is the basis for the scheme proposed by Charles Bennett and his collaborators, uses a random encryption key that is as long as the message itself. The sender of the message, Alice, starts by digitizing the text to be sent to the recipient, Bob, by translating it into a sequence of zeros and ones (the binary language of computers). To this sequence she adds the key, which consists of another sequence of zeros and ones that is as long as the original message. She transmits the result to Bob, who has the key that Alice used. Bob subtracts the key and what is left is the original message. To read it, it will of course be necessary to convert the series of zeros and ones into a sequence of letters, but this is routine work for any computer. To make this system truly private it is essential that the key is a random sequence and that it is only used once. This means that these numbers have to be generated in advance and that Alice has to get them to Bob. And that is where the problems begin. If Alice and Bob never meet with each other personally, as is normally the case with the people involved in an e-commerce transaction, then they will have to place their trust in the method used to transmit the key. And how is this to be done? In an encrypted form? But for that they need to have agreed on another key, so the problem seems to be impossible to solve. At some point, Alice and Bob will have to meet or to entrust the information to a messenger. But as the meddling Eve is always on the prowl, they know they will never have absolute security.

This is where the world of quantum physics comes in. This world has strange rules that are impossible to realize intuitively on the basis of our daily experiences. One of them is uncertainty. And this uncertainty is not based on our lack of knowledge, it is an intrinsic property of subatomic particles.

Alice starts by sending Bob a sequence of light particles, in other words, a series of photons. Her device contains two polarizing filters, one that is oriented vertically and another that is at an angle of 45°, as shown in the illustration. To create the key, the device alternates the polarizing filters randomly, for example assigning the number 0 to a photon that is polarized vertically and the number 1 to a 45° polarized photon. In Bob's device there are also two polarizing filters, one positioned horizontally and another at an angle of −45°. When the device receives each photon it passes it in completely random fashion through one of the filters.

The photons sent by Alice and received by Bob may or may not pass through his device depending on the combination of polarizing filters. If Alice sends a vertically polarized photon and Bob's device makes it pass through the horizontal filter, then the photon will be retained and will not pass through the device. If Alice sends a 45° polarized photon and Bob's device passes it through the −45° filter, then the particle will also be retained and will also not pass through. Polarizing filters that are perpendicular to the direction of polarization retain the particles.

Surprising things happen, though, when Alice sends a vertically polarized photon and Bob's device passes it through the diagonal filter, or when Alice sends a diagonally polarized photon and Bob's device passes it through the horizontal filter, that is when the difference in the direction of polarization between the filters in the two devices is a 45° angle. In this case quantum uncertainty enters the field: half of the particles pass through Bob's filter, but the others are retained by it, and it is impossible to know in advance which particles will pass through and which will be retained.

At this point only Alice knows the polarization of the photons that were sent, while only Bob knows which photons made it to their destination. Therefore Bob gets to know the polarization that Alice used for the photons that passed through the filters, as a photon that passed through his diagonal filter must have been polarized vertically by Alice, which means it has the value 0. A photon that passed through his horizontal filter must have been polarized diagonally by Alice, so it has the value 1. However, Bob knows nothing about the particles that were retained.

Alice now needs to know which photons passed through the filters in Bob's device. For this purpose she can contact Bob by means of any channel whatsoever, however unsecured. This conversation can even be overheard by Eve, as even if she knows which photons were received by Bob's device, she will not know which filter they each passed through. So now the key has been set up using only the photons that completed their journey and Alice and Bob can communicate in absolute security. It is the quantum physics uncertainty principle that gives them the certainty that nobody is eavesdropping.

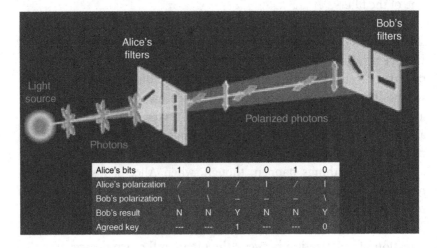

Alice's bits	1	0	1	0	1	0
Alice's polarization	/	I	/	I	/	I
Bob's polarization	\	\	–	–	–	\
Bob's result	N	N	Y	N	N	Y
Agreed key	---	---	1	---	---	0

Only a few years ago all this would have sounded like science fiction, but recent technological advances have thrust it into the realm of possibility. As you can imagine, there are immense technical difficulties still to confront. For example, how can you emit light photon by photon? How do you ensure that these particles reach their destination? One by one these problems have been solved. Scientists have succeeded in putting quantum cryptography into practice via optical cables and via air for several miles. Maybe we are not far away from being able to protect our secrets by making them travel particle by particle. At the speed of light.

THE FBI WAVELET

The language of mathematics can seem esoteric and purely abstract, but many of its constructions end up having surprising applications. One of the most recent and spectacular successes of mathematics is taking place in the processing of signals, and in particular in the processing of images. This new technique has a quaint name: wavelet analysis.

As always, this new tool didn't fall from the sky: its origins can be traced to the work of the French mathematician Jean Baptiste Joseph Fourier (1768–1830), who created a technique known as "harmonic analysis", or more commonly, Fourier analysis. Jean Baptiste studied at the military academy in Auxerre, the city of his birth, where he stayed, drawn to mathematics. He later entered a seminary with the objective of becoming a monk, but then the French Revolution intervened. Gradually he began to support the revolutionary movement, and eventually declared himself ready to fight for a free government, liberated from kings and priests. In the turmoil of the revolution he was taken prisoner and nearly condemned to death. Subsequently, he lectured in Paris and accompanied Napoleon on his Egyptian campaign as a member of his scientific committee. When he returned to France, he was appointed prefect of the Isère department, worked in the institute of statistics, and was elected a member of the Academy of Science in Paris.

Fourier's most significant scientific treatise, *La Théorie Analytique de la Chaleur (The Analytical Theory of Heat)* is one of the scientific milestones of the 19th century. Fourier surprised his contemporaries by contending that functions could be represented as the sums of waves, that is, by the sums of the known trigonometrical functions sine and

N. Crato, *Figuring It Out*, DOI 10.1007/978-3-642-04833-3_23,
© Springer-Verlag Berlin Heidelberg 2010

cosine, the *sinusoidal waves*. He believed it was often easier to mathematically process the sums of waves than their original functions, and that the sums are equivalent to the initial functions, provided a sufficient (possibly infinite) number of constituents are added. It has turned out that this seemingly odd idea is very versatile. Mathematicians have come to rely upon Fourier analysis to resolve many problems that would have been impossible to process by any other means.

In 1965 the work performed by James Cooley and John Tukey at the Bell Laboratories in New Jersey transformed Fourier's idea into an extremely practical technique. Cooley and Tukey created a new algorithm to calculate the Fourier series, calling it the *Fast Fourier Transform* (FFT). Today it is used in a wide variety of areas, from the analysis of radio signals or econometric forecasting to clinical studies of brain waves. The mathematician Gilbert Strang, from MIT, observed that "whole industries are changed from slow to fast by this one idea – which is pure mathematics".[1]

More recently, in the 1970s and 1980s, various engineers and mathematicians began to attempt to resolve certain practical limitations of the Fourier series. For instance, if we want to codify one of Bartok's concerts using sinusoidal waves, then we need a huge number of such waves because the concert has brusque changes, whereas the waves continue indefinitely. Mathematically, this means that a gigantic number of coefficients is required to codify it properly. This is not at all practical for the purposes of a digital recording.

Fourier's technique, however, is perfect for capturing one specific moment of the concert. To recreate the original sound you only have to synthesize the different frequencies, the different notes, and reproduce the timbres of the different instruments, which are added to these notes. But because the notes are in constant flux, using the waves to capture the entire concert is not a practical proposition. The idea that some people had was to create mini-waves with a precise beginning, middle and end, and to use those mini-waves to analyze the original signal. Yves Meyer, at

---

[1] Wavelet transforms versus Fourier transforms, *Bulletin of the American Mathematical Society*, 28–2, April 1993, 288–305.

the École Polytechnique, and Ingrid Daubeschies, a Belgian mathematician who was working at Bell Laboratories at the time, played a major role in the development of this idea. A new mathematical tool had been born.

In French the new functions were called *ondelettes*, or small waves. So in English they became known as *wavelets*.

When the FBI consulted a group of mathematicians on how best to process the enormous archive of fingerprints held by the federal agency, they proposed using wavelets. The FBI had been storing fingerprints since 1924. By 1996 their archive contained 200 million files, and it continues to grow at a rate of almost 50,000 new files every day.

When the FBI had earlier begun transmitting images from the archive electronically, they had observed that the system they were using was very slow, and that images had to be compressed in order to be properly transmitted. JPEG, the most commonly used compression system on the internet, produced a very grainy image. When a significant compression of the image was required, the details disappeared and the sharp transitions between lines were blurred.

After a great deal of mathematical processing and many experiments by the consulting mathematicians, the FBI decided to switch to a wavelet-based system for compressing the images. They created a new wavelet that was uniquely suitable for reproducing fingerprints, which made it possible to achieve significant reductions in the size of the electronic files. Another mathematical achievement.

The Enigma Machine

Since ancient times men have dreamed of constructing an automatic coding machine. As far as we know, the first attempt was made by the Renaissance architect Leon Battista Alberti (1404–1472), who positioned one concentric disk on top of another, each one inscribed with all the letters of the alphabet. By turning one disk to a certain position with respect to the other, he was able to pair each letter with another, which served to automate the task of encrypting messages by substituting letters. The device may only have mechanically reproduced actions that could be done mentally, but it did ensure that no substitution errors were made in the process.

The messages obtained by this simple process can be easily decrypted by statistical analysis. In English, for example, the most frequently occurring letter is E (with a mean frequency of 12.7%), followed by T (9.1%) and A (8.2%), with Z at the end of the list (0.1%). If the encrypted message is sufficiently long, it is relatively simple to identify some of its letters. Then, half the job is done. After some letters and parts of a message have been decrypted, it is usually simpler to complete the task.

In 1918 the German inventor Arthur Scherbius (1878–1929) developed an automatic coding machine that used disks similar to those Alberti had constructed but also featured some improvements. It resembled a typewriter and it used electrical connections, transforming anything typed on the keyboard into an encrypted message that was displayed in illuminated letters. Three disks in a series, usually called rotors, formed the core of his invention. The first disk translated the

N. Crato, *Figuring It Out*, DOI 10.1007/978-3-642-04833-3_24,
© Springer-Verlag Berlin Heidelberg 2010

original letter into a coded letter to be transformed by the second disk into yet another letter, which in turn was transformed by the third disk. Scherbius' decisive improvement was to rotate the position of the disks. After the first disk had rotated through all of its positions, the second disk then rotated one position. After the second disk had rotated through all of its positions, the third disk rotated one position. By this means, Sherbius ensured that any letter's encryption would be repeated only after a complete cycle of all three disks. Using our alphabet of 26 letters, this means a cycle of $26 \times 26 \times 26 = 17,576$ positions before the code is repeated, which is a sufficiently large number to prevent the use of statistical analysis in discovering the frequencies.

Each of the disks in the machine invented by Scherbius had its own corresponding letters. For example, if one of them transformed A into S and B into F, etc., then another might transform A into H, B into Z, and so on. By rotating the removable and permutable disks, his machine multiplied the number of possible transformations by $3 \times 2 \times 1 = 6$. Finally it had a series of electrical connections that exchanged six of the letters each time. Now, the number of possible ways of connecting six pairs of letters in the alphabet is gigantic, more than one hundred thousand million. All in all, Scherbius' machine allowed about ten thousand trillion codes to be created. Scherbius called his invention the Enigma machine.

Initially Scherbius' invention was a commercial disaster. However, a few years later, the potential of the Enigma machine was recognized when the German armed forces were beginning to make preparations to launch WWII. The German military ordered thousands of the machines, convinced that by using it, their encrypted military secrets would be kept completely safe. In this they were mistaken, as a Polish statistician and an English mathematician would discover a way to break the Enigma code.

Marian Rejewski was 23 years old when he was summoned to work with the *Biuro Szyfrów*, the Polish General Staff's Cipher Bureau. His recent university training in mathematics and statistics was tailor-made for the Bureau's new recruitment policy, which was to concentrate on hiring young mathematicians. Rejewski (1905–1980) proved to be a real catch, as he developed pioneering techniques that had been thought impossible until then. By the early 1930s, at a time when the British and French were despairing of ever succeeding in breaking the German code, the Poles were able to decrypt the German messages in a matter of hours. But in December, 1938, improvements were made to the Enigma machine which made breaking the German code more difficult: an additional two rotors brought the total up to five, from which three

Alan Turing (1912–1954) and his colleagues at Bletchley Park in southern England knew that the German messages had to contain some inevitable words, such as "weather", "attack" or "ship". The search for sequences that might contain such words, called *cribs*, was one of the techniques most often used to decrypt messages. The film *Enigma*, based on the book of the same name by Robert Harris, included some very emotional scenes of code-breakers searching for *cribs*. If you want to understand the mode of operation of the Enigma machine and the story of how it was broken, the standard reference source is David Kahn's book, *Seizing the Enigma* (published by Arrow, London, 1966). Another remarkable book that devotes many pages to Enigma and includes a clear explanation of the principles used by the machine and by the code-breakers is *The Code Book*, by Simon Singh. If you are interested in learning how the Enigma machine works, you can also look on the internet at the website www.bletchleycovers.com, where you will find many links to many other sites as well as a virtual Enigma machine where you can type in messages and watch as they appear in an encrypted version were selected to be used, and the number of electrical connections was expanded from six to ten. Rejewski had proved that the German system was not invincible, but now the task had become too complex for the computational means at his disposal. The Poles, who assumed that they were soon going to be invaded by the Germans, decided to reveal their knowledge of the Enigma machine to the British and French. On August 16, 1939, two replicas of the machine and two sets of decryption plans were sent to London and Paris. Two weeks later, on September 1, Hitler invaded Poland and the war officially began.

The British foresaw the importance of cryptography, and they assembled an exceptional group of experts at Bletchley Park, a stately home 50 miles to the northwest of London. One of those gifted cryptologists was the mathematician Alan Turing (1912–1954), who would eventually become the main protagonist in breaking the Enigma code. Inspired by his previous studies on a theoretical computing machine, Turing detected various weak points in the German system and invented machines (which were called "bombes") that could automatically reproduce the sequences of the rotors in the Enigma machine. His success would play a decisive part in the victory of the Allies. Churchill called the cryptologists at Bletchley Park "the geese that laid the golden eggs but never cackled".[1]

---

[1] http://infosecurity.us/?p=5735

ART AND GEOMETRY

THE VITRUVIAN MAN

"The Vitruvian Man!" Langdon gasped. Saunière had created a life-sized replica of Leonardo da Vinci's most famous sketch.

This is one of the introductory scenes in *The Da Vinci Code*.[1] The museum curator Saunière had drawn a circle around himself and painted a five-pointed star in blood on his stomach. In the following 500 pages Professor Langdon will provide an explanation of the geometry of the star and the figure of the Vitruvian Man. The book is a mixture of facts and fiction, which is perfectly acceptable as it is a novel. But as a reader you are entitled to know which of the book's facts have not been embellished.

The five-pointed star, also known as a *pentagram*, is a geometrical figure that has been a subject of investigation and curiosity for thousands of years. It has represented a mystical symbol for various civilizations, in some cases being associated with the supreme divinity, and in others, with the Devil. Placed on its "feet" it resembles the human figure. With one point at the bottom and two points at the top it looks like a horned animal, naturally assumed to be of a diabolical nature. In this magical association it is often called a *pentacle*.

The ancient Greeks studied the geometry of the pentagram. The mathematician Euclid (fl. 300 BC) showed how to draw it with a non-graduated ruler and a pair of compasses, which gave it a classical geometrical dignity. He also revealed some of its curious properties.

---

[1] P. 45 of the first American edition, DoubleDay, New York, 2003.

N. Crato, *Figuring It Out*, DOI 10.1007/978-3-642-04833-3_25,
© Springer-Verlag Berlin Heidelberg 2010

If you join one point of the star to the two points opposite, you obtain an isosceles triangle. This triangle has two angles of 72° and a third angle of 36°, which is half of either of the larger angles. Such a polygon is called a golden triangle. Curiously, if we bisect one of the larger angles, thus dividing the original triangle into two, the smaller resulting triangle is similar to the original triangle in that it is also a golden triangle. If we divide this angle using the same procedure, we can construct an infinite series of golden triangles, one inside the other.

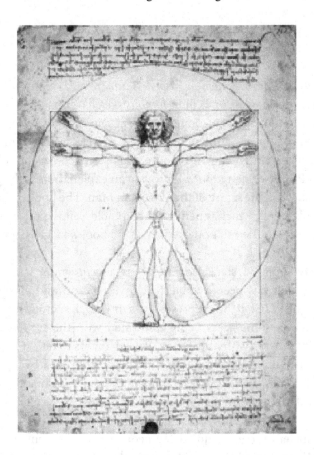

By joining the points of the pentagram we can draw another curious geometrical structure, a regular pentagon that contains the star. Looking at its center we discern another regular pentagon. This means that we

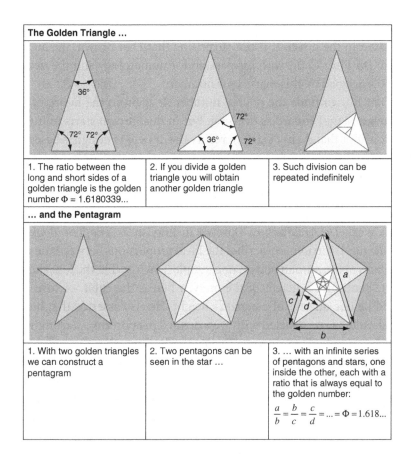

The Golden Triangle ...

1. The ratio between the long and short sides of a golden triangle is the golden number Φ = 1.6180339...	2. If you divide a golden triangle you will obtain another golden triangle	3. Such division can be repeated indefinitely

... and the Pentagram

1. With two golden triangles we can construct a pentagram	2. Two pentagons can be seen in the star ...	3. ... with an infinite series of pentagons and stars, one inside the other, each with a ratio that is always equal to the golden number: $$\frac{a}{b} = \frac{b}{c} = \frac{c}{d} = ... = \Phi = 1.618...$$

can construct an infinite series of pentagons and pentagrams, one inside the other.

The ratio between the shorter and longer sides of a golden triangle is an irrational number that is approximately equal to 1.618. This means that the distance from one of the five points of the star to one of the opposite points is equal to this number times the distance between two contiguous points. And that is not the end of the surprises: the ratio between this latter distance and the length of a segment of the star is also this mysterious number. You may already have guessed that we are talking about the golden number or Φ (phi), which is referred to so many times in Dan Brown's novel. This is the same number that appears as the limit of the ratios between successive terms in a Fibonacci sequence, also discussed in the novel. It is not surprising that Dan Brown, the author of *The Da Vinci Code,* with his passion for numerology and the occult, was so intrigued by this marvelous number.

Dan Brown finds the golden number Φ again in the figure of the Vitruvian Man, Leonardo's drawing. But in this, Brown starts to invent relationships that do not exist. In reality the drawing's design is based on the human body's simple proportions, proportions that are expressed as integers, not as irrational numbers like Φ.

In creating the drawing, Leonardo followed the instructions of the Roman architect, Marcus Lucius Vitruvius Pollio (about 90–20 BC). In his work *De Architectura* (known in English as *The Ten Books of Architecture*), Vitruvius describes the ideal proportions between the various parts of the human body. For example he wrote that the foot should be one sixth of the height of the body, whereas the cubit, or forearm, should be one quarter of a person's height. He also asserted that buildings should be constructed with well-defined proportions, starting with a "perfect number". (Subsequently he questions whether this number is six, considered the first perfect number, or ten, the number which "Plato held as the perfect number". In defense of the number six, he observes that mathematicians "have said that the perfect number is six, because this number is composed of integral parts which are suited numerically to their method of reckoning: thus, one is one sixth; two is one third; three is one half; four is two thirds, five is five sixths, and six is the perfect number" [2]).

No traces of the golden number can be discerned in the works of Vitruvius. Despite a very widespread but mistaken view, even in academic studies, it does not appear in Leonardo's drawing either. Why don't you take the trouble to accurately measure the ratio between the circumference of the circle and the side of the square in the drawing of the Vitruvian Man? By using a simple ruler and tape measure you can lay to rest many claims in the realm of numbers . . .

---

[2] Morris Hickey Morgan translation of *De Architectura* in http://www.perseus.tufts.edu/cgi-bin/ptext?doc=Perseus%3Atext%3A1999.02.0073;query=chapter%3D%2321;layout=;loc=3.preface%201

THE GOLDEN NUMBER

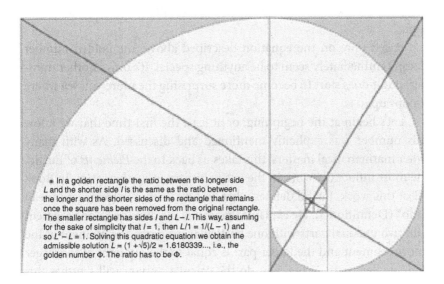

* In a golden rectangle the ratio between the longer side L and the shorter side *l* is the same as the ratio between the longer and the shorter sides of the rectangle that remains once the square has been removed from the original rectangle. The smaller rectangle has sides *l* and L−*l*. This way, assuming for the sake of simplicity that *l* = 1, then L/1 = 1/(L − 1) and so $L^2 - L = 1$. Solving this quadratic equation we obtain the admissible solution L = (1 +√5)/2 = 1.6180339..., i.e., the golden number Φ. The ratio has to be Φ.

There are numbers that surprise us. They pop up unexpectedly in all sorts of situations. For example, take π, the number that represents the quotient of the perimeter of a circumference by its diameter. This number also appears in equations representing the area of a circle, as well as the surface and volume of a sphere. It is not difficult to accept this, as the circumference must have something to do with these other measurements. But it is not so easy to understand the reason why π also appears

N. Crato, *Figuring It Out*, DOI 10.1007/978-3-642-04833-3_26,
© Springer-Verlag Berlin Heidelberg 2010

in statistics, in complex exponential functions and even in the sum of numerical series[1] such as $1 + 1/4 + 1/9 + 1/16\ldots$

Another of these surprising numbers is the golden number, also known as the golden ratio, or divine proportion. The golden number is normally represented by the Greek capital letter Φ (phi), and is equivalent to half the sum of one plus the square root of five. It is an irrational number, given as the infinite non-periodic decimal $1.61803398\ldots$

Mario Livio, an American astronomer at the Hubble Space Telescope Science Institute, has published a book in which he calls the golden number the most surprising number in the world (*The Golden Ratio*, Review, London). It may seem astounding that somebody could write an entire book whose only subject is this strange number, but really it is only the latest of several books and innumerable articles on this topic.

Based only on the equation described above, the golden number doesn't immediately seem to be anything special. It's only another number. But it does start to become more surprising the more you see where it pops up.

Let's begin at the beginning, or at least the first time that we know this number was explicitly mentioned and discussed. As with many other mathematical matters, this takes us back to the *Elements* of Euclid, the most influential work in the entire history of mathematics. In Book VI of this work, Euclid defines what he calls "cut in extreme and mean ratio" (Definition 3). He explains that this is the division of one segment into two unequal parts with one particular property: the quotient of the entire segment and the larger part is equal to the quotient of the larger and smaller parts. When you do the calculation you will see that this ratio has to correspond precisely to Φ, the golden number.

This subject was referred to again many times, notably in the 13th century by Leonardo of Pisa (about 1170–1240), better known as Fibonacci, and by Fra Luca Pacioli (1445–1517), who first introduced the expression "divine proportion". It was only in the middle of the 19th

---

[1] In fact, the sum of the series of the inverse squares here initiated is a fourth of the square of pi.

century that terms such as "golden ratio" and "golden number" were coined. The letter phi (Φ) as a symbol for this number only appeared at the beginning of the 20th century.

The number Φ appears in many geometrical constructions. For instance, let's take the isosceles triangle (two equal angles) in which the smaller angle is half of either of the two larger (equal) angles, i.e., one angle is 36° and the other two are each 72°. In this instance, the golden number appears as the ratio between the long and short sides of the triangle. In addition, if we divide one of the larger angles in half, we obtain two triangles, the smaller of which is similar to the triangle that we divided — the sides have the same ratio as the original triangle and the angles are identical.

However, the most celebrated geometrical construction of all is the so called "golden rectangle", in which the ratio of the rectangle's sides is the number Φ.

There has been much speculation on the aesthetic properties of this golden rectangle. There are those who maintain that the proportions are so perfect that they were incorporated into such ancient architectural masterpieces as the façade of the Parthenon in Athens. Likewise, there are those who claim that the golden number appears in the Great Pyramid in Egypt, as the ratio between the height of a lateral triangle and half of its base. As the mathematician George Markov, from the University of Maine (www.umcs.maine.edu/~markov) recently demonstrated, though, these speculations are not based on reality. What happens is that, as all these monuments have so many possible measurements that can be compared, it only takes several attempts to find an approximation of one interesting number or another. But what is not in doubt is that the golden rectangle is especially beautiful and possesses an aesthetically appealing appearance. In some studies where persons were requested to select one of several rectangles as the most beautiful, many of those surveyed did in fact choose the golden rectangle. This geometrical figure seems to be more visually pleasing than, for example, the rectangle of a sheet of A4 paper.

However, both these rectangles can be divided successively into figures that are always similar. In the A4 rectangle, the division is in half, creating two rectangles with the ratio of the sides being identical to those of the original rectangle.

A different method is required if you want to successively divide a golden rectangle to produce similar rectangles. The original golden rectangle is divided in such a way that a square and a rectangle are obtained. There is only one way to do this, which is to create a square with sides that are equal to the shorter side of the original rectangle. It just so happens that the remainder is another golden rectangle.

If you divide a golden rectangle successively in accordance with this rule, you will obtain ever smaller rectangles, one inside the other. We can draw a spiral in them that converges on a point (a "pole") that is at the intersection of two diagonals: that of the original rectangle and that of the golden rectangle created by the first subdivision. The spiral drawn in a succession of golden rectangles is called a "logarithmic spiral" and is to be found in the most varied situations, including the shells of marine animals, the flight trajectory of falcons, flowers and galactic spirals.

One of the most amazing examples of the golden number can be found in the arrangement of the petals of a rose. They are separated at an angle that is a fraction of Φ. This arrangement permits the positioning of the petals in a compact form and simultaneously maximizes their exposure to light. Just like mathematicians, nature too seems to be eternally fascinated by the golden number.

THE GEOMETRY OF A4 PAPER SIZES

The paper format generally used in photocopiers and printers everywhere outside North America, and which is also generally used for letters and writing pads, has the curious name of A4. Measuring 210×297 mm (approximately $8\frac{1}{4} \times 11\frac{3}{4}$ in.), A4 sheets are an unusual size; it would certainly seem more logical if this measurement were a round number. Why not 200×300 mm, for example?

It turns out these bizarre dimensions are actually the result of a carefully considered convention, which has now been adopted by most countries in the world, and is based on the German DIN system (**D**eutsches **I**nstitut für **N**ormung e. V.). The German standard was adopted as the ISO 216 standard by the International Organization for Standardization. In accordance with this standard, there is a series of basic paper sizes, starting with A0, the largest, and then decreasing in size to A1, A2, A3..., and ending in the minuscule A10, which measures only 26×37 mm (about $1 \times 1\frac{1}{2}$ in.). All these sizes are designed in such a way that the size with the next highest number can be obtained by folding the sheet in half. To take an example: if you fold an A0 sheet in half, you have an A1 size, or folding an A4 sheet in half gives you an A5. But there is a lot more to this standard than that. The sizes are designed so that they will always maintain the same ratio between the length and width of the paper. The sizes are rounded off to the nearest millimeter (about one twenty-fifth of an inch), which is quite a reasonable approximation.

This rule has many practical applications, especially in photocopiers. If you place two A4 sheets side by side on the photocopier and

N. Crato, *Figuring It Out*, DOI 10.1007/978-3-642-04833-3_27,
© Springer-Verlag Berlin Heidelberg 2010

select the reduced size mode, it is possible to photocopy each of the two original A4 sheets exactly on one half of the A4 copy produced by the machine. It is easy to see that not every paper format would allow you to do this. For instance, if the original sheets were square, and we wanted to copy two of them, in reduced size mode, on to a single square sheet, we would have to waste half of the photocopy. But with the A4 system there is no waste, as the ratio is maintained when we fold the sheet in half.

What format does a sheet of paper have to have in order for it to maintain its ratio when it is divided in two? A few simple calculations will reveal that. The sides of the rectangle have to have a ratio of one to the square root of two (approximately 1.4142). There is no other solution. If you do the math, you will agree that $210 \times 1.4142 = 296.982$, or for all practical purposes 297. Those are the proportions of the A4 series.

All this follows a perfectly logical system, but we still have to define our starting point. How was the A0 size designed? Curiously enough, this was not an arbitrary decision either. It was defined with the ratio of the length and width set at one to the square root of two, as usual in this system, but with the added restriction that the A0 size has an area that is equivalent to one square meter (about 10.75 square feet). That completed the definition of the A4 system. Luckily, the result was an A4 paper size that is excellent for office work.

This standard also makes it simple to calculate the weight of a ream of paper, for example. The weight of the paper, or "grammage" as the professionals say, is calculated in grams per square meter. Normally we might use paper that weighs 80 g/m². This means that one A0 sheet of this paper weighs 80 g. As one A0 sheet has the same surface area as 16 A4 sheets, each A4 sheet weighs 5 g and the whole ream (500 sheets) of A4 paper weighs 2500 g or 2.5 kg (approximately 5 pounds 8 ounces).

The system defined by the ISO 216 standard also encompasses two other series of sizes. These are the B series, used for envelopes that contain the equivalent sheets of paper from the A series, and the C series, used for slightly smaller envelopes that may contain fewer sheets of paper. For example, if you need an envelope to mail an A4 brochure, you can use a B4 envelope, which measures 250×353 mm (almost

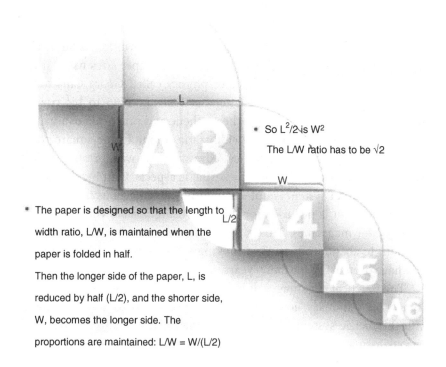

* So $L^2/2$ is W^2

The L/W ratio has to be $\sqrt{2}$

* The paper is designed so that the length to width ratio, L/W, is maintained when the paper is folded in half.

Then the longer side of the paper, L, is reduced by half (L/2), and the shorter side, W, becomes the longer side. The proportions are maintained: $L/W = W/(L/2)$

10×14 in.). If you want to mail a thin A4 document, you could use a C4 envelope, which measures 229×324 mm (about $9 \times 12^3/_4$ in.). This makes it simple for retailers to know what their customers need.

But the exact dimensions of the envelopes are also mathematically logical. The geometrical mean between two consecutive A sizes was used to define the dimensions of the B series. For example, the geometrical mean between the dimensions of A4 and A3 paper was used to calculate the dimensions of a B4 envelope. A similar procedure was also used to calculate the C series, so that the C4 envelope is defined by the geometrical mean between the A4 and B4 sizes.

The geometrical mean is a mean, as a pedant would state, that results in intermediate dimensions between extreme values. But it is a special mean. It is obtained from the square root of the products of two values. That is why it maintains the relative proportions. So B4 is to A4 as A3 is to B4.

This whole complex system evolved over a period of two hundred years, and finally it was adopted almost all over the world. As far as we know, the first person to think of standardizing paper sizes using similar rules was a German professor of physics called Georg Christoph Lichtenberg (1742–1799). In a letter he wrote in 1786 to his friend Johann Beckmann, he described the aesthetic and practical advantages of using paper with a length to width ratio of one to the square root of two. With regard to the practical advantages he was certainly right, but opinions are divided about the aesthetic aspects. Graphic designers know that the A4 system is not aesthetically advantageous for placards or magazines—it is not used for the advertising posters you see in the streets or for magazines. This is one of the reasons why Americans do not want to abandon their good-looking *letter* format (8 × 11 in.) that they have been using for many years.

The practical advantages of Lichtenberg's method were seen by the French revolutionary government when it decided to adopt it along with the metric system. In 1794 the French *Loi sur le Timbre* defined various formats equivalent to the current ISO standard sizes. The formats defined were called *grand registre* (currently A2), *grand papier* (B3), *moyen papier* (A3), *petit papier* (B4), *demi feuille* (B5) and *effets de commerce* (1/2 B5). Only the A4 size was missing.

Today the ISO standard is used in most parts of the world except in North America. So the next time you pick up a sheet of A4 paper, just remember that you are holding an object with a remarkable mathematical history.

The Strange Worlds of Escher

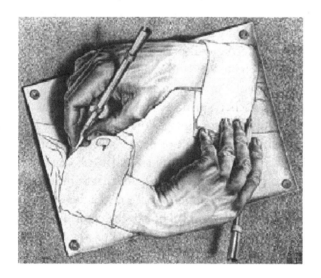

Maurits Cornelis Escher was born in 1898, in the city of Leeuwarden in the Netherlands. During his life he produced the most intriguing and mathematically sophisticated woodcuts any artist has ever created.

The young Maurits had an ordinary childhood. His grades at school were average and he showed little interest in his studies, but under pressure from his family he agreed to study architecture at the Haarlem School of Architecture and Decorative Arts, where he met a graphic arts teacher who would transform his life. This teacher, a Dutch Jew of Portuguese descent named Samuel Jessurun de Mesquita, taught his

N. Crato, *Figuring It Out*, DOI 10.1007/978-3-642-04833-3_28,
© Springer-Verlag Berlin Heidelberg 2010

student design techniques and awakened his interest in lithographs and woodcuts. Maurits Escher soon gave up his study of architecture and switched to graphic art, where he was tutored by Jessurun de Mesquita.

Once he had completed his studies, Escher decided to see the world. He spent time in Spain and then Italy, where he met his future wife, and moved to Rome. He was always able to pay his way by working as a graphic artist. However, when the political climate in Italy became unbearable under Mussolini, he moved to Switzerland, and later to Belgium. In 1941 he returned to the Netherlands, where he spent the rest of his life.

On one of Escher's trips to Spain he visited Granada. He was entranced by the Moorish tiles and the intricate patterns that are typical of Arab art. This experience fed his passion for geometric grids, the division of a plane into geometrical figures, with a motif that repeats itself, reflects itself, becomes displaced and rotates.

Escher was fascinated by the challenges and limitations of two-dimensional representation, which is clearly evident in nearly all of his work. One of Escher's principal preoccupations was the creation of illusory three-dimensional images on plane surfaces. But above all, he was intrigued by the conflict of perspectives. The lithograph entitled *Drawing Hands*, which he produced in 1948, depicts a hand that is drawing a hand that is drawing a hand that is drawing a hand ... The hands belong to a two-dimensional world and also simultaneously to a spatial world. They pass from one to the other. Escher argued that we are trained to see either a plane surface or a volume in an image. Either one or the other. He found it interesting to provoke a conflict of representations.

But this lithograph is also interesting for another reason. It illuminates one of the fundamental problems of modern times. In *The First Moderns* (Chicago, 1997), a study of the scientific and artistic revolutions of the last 100 years, author William Everdell posits that the "barber paradox", attributed to the mathematician, logician and philosopher Bertrand Russell (1872–1970), is one of the milestones of the early 20th century. The paradox consists of the question "Who shaves the barber who shaves all but only those who do not shave themselves?" Whatever you reply, you contradict the terms of the question. Using similar paradoxes, Russell effectively questioned the attempts by Gottlob Frege (1848–1925) and generations of mathematicians before and after him to construct a perfect, self-referencing logic. Using similar arguments, Kurt Gödel (1906–1978) showed that the endeavors of generations of mathematicians at the turn of the century, notably David Hilbert (1862–1943), to base mathematics on an eternally perfect and once-and-for-all complete logic were bound to fail. A hand drawing a hand drawing a hand cannot be a real hand.

Relativity, 1953

Another interesting example of the strange world of Escher is provided in *Relativity,* a lithograph produced in 1953. It contains three completely different worlds that are combined within one image. In each one the perspective is coherent, creating its own world. But the ceiling of one world is the wall of another. At the junction of two worlds a door to one is also a trapdoor to another. Two of the central staircases can be used from either side (both upside-up and upside-down), with one world on one side and another one on the other.

Each of these worlds has its own field of gravity, each of them pulling in their own direction. One of the most interesting surprise effects of this lithograph can be found in the persons who seem to be strolling through worlds that are not their own. The man carrying a sack in the center of the picture belongs to the world on the left, where gravity moves objects from left to right. He appears to be walking through other worlds, yet, for him, everything is coherent and he will be able to reach the garden on the upper right edge of the picture by simply turning right and then going up the stairs. On these stairs, the ones at the top of the image, are two figures from two different worlds.

Bruno Ernst, who has studied the work of Escher, has suggested that it might be useful for astronauts in gravity-free situations to contemplate this image, as they need to become accustomed to using any plane of reference in their compartments. They can even pass each other moving in perpendicular directions. In space they might encounter the strange worlds of Escher.

Escher and the Möbius Strip

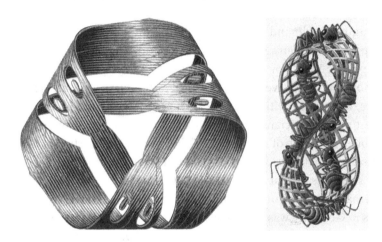

Escher was once quoted as saying: "In 1960 I was exhorted by an English mathematician (whose name I do not call to mind) to make a print of a Möbius strip. At that time I scarcely knew what it was".[1] He responded to this challenge by producing two images that became famous: *Möbius Strip I* and *Möbius Strip II*, which I've reproduced here. In the first of these woodcuts, which seems to depict three snakes biting each others' tails, Escher invites us to follow the line of the snakes. What we discover, to our surprise, is that the three reptiles are all on the same surface

---

[1] Bruno Ernst, *The Magic Mirrors of M.C. Escher*, Taschen America, 1994, page 99.

N. Crato, *Figuring It Out*, DOI 10.1007/978-3-642-04833-3_29,
© Springer-Verlag Berlin Heidelberg 2010

even though they appear to be following two distinct orbits. In the second woodcut, *Möbius Strip II,* we see nine ants all crawling in the same direction. This time Escher asks us to follow their path and confirm that it is indeed a path without end, because no matter which starting point you choose, you always end up at the same point. The ants appear to be crawling on two separate sides of a single surface, but ultimately each of them travels the entire length of the surface on which they are crawling. In both these images the paths are endless.

As we've noted, these two woodcuts are based on what we today call a Möbius strip, named after its discoverer, a German mathematician and astronomer who performed important work in geometry, topology and complex analysis. The woodcut with the ants shows an actual Möbius strip.

August Ferdinand Möbius was born on November 7, 1790, in Schulpforta, Germany, near both Leipzig and Jena in Saxony. He died in Leipzig in 1868. For nearly his entire adult life he was a professor of astronomy at the University of Leipzig. At the start of his mathematical studies, he worked with Carl Friedrich Gauss (1777–1855), one of the greatest mathematicians who ever lived, and the only one up to now to be dubbed a "prince of mathematics". He completed his studies by writing his doctoral thesis on the occultation of fixed stars, and then, like so many scientists at that time, he dedicated his life to the study of both mathematics and astronomy. In the course of his life the field of mathematics in Germany was completely transformed. At the time of his birth it would have been difficult to identify even two German mathematicians of international renown, but by the time he died, Germany had become one of the main centers for the teaching of and further research into mathematical concepts, which gave it a powerful influence on scientific matters throughout the world. Möbius was a key participant in this extraordinary development, which is not surprising given the political and social changes that were then taking place in Germany. Of course, it was during this period that Germany ceased to be a conglomeration of small states, and became an empire under the aegis of Prussia.

Among mathematicians, Möbius is known for many results and constructions, principally the transformation (function) that is named

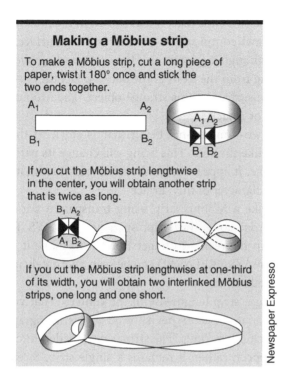

Making a Möbius strip

To make a Möbius strip, cut a long piece of paper, twist it 180° once and stick the two ends together.

If you cut the Möbius strip lengthwise in the center, you will obtain another strip that is twice as long.

If you cut the Möbius strip lengthwise at one-third of its width, you will obtain two interlinked Möbius strips, one long and one short.

Newspaper Expresso

after him, and which continues to play an important role in complex analysis. He is also known for various contributions to geometry and topology, a branch of mathematics that is, from many points of view, a generalization of geometry.

In his topological studies Möbius was especially interested in one particular property of surfaces, that of the possibility or impossibility of orientation. For this study, he constructed a non-orientable surface that became known as a Möbius strip. To make it he literally had to change sides.

A sheet of paper has two sides (front and back), as well as a single edge that runs around the corners of the page. Can it be that a sheet of paper could have a single side and a single edge, so that an ant could travel from one side to the other without crossing that edge? Today we know that this is indeed possible: all you need to do is twist the paper

180° and stick the two ends together, as shown in the illustration above. The ants can crawl continually over the resulting surface, which is a Möbius strip. Although it appears as if they are crawling over both sides of the strip, going from the front to the back, in fact they are only moving on the single surface of this strange object. The Möbius strip does not have a front or a back.

Imagine a two-dimensional being that is stuck to the strip, not crawling over it like the ants. This being will change its parity when the strip is rotated, i.e. it appears as a mirror image, a fact that reflects the non-orientable character of this surface.

You can make a Möbius strip using transparent paper and then write some words along the strip, such as ALWAYS PREPARED. Holding the strip between two fingers, pull it to make it rotate completely. The ALWAYS PREPARED can be seen alternately in its normal and in its inverted form.

Cut all along a strip lengthwise in the center. What you obtain is somewhat similar to what Escher showed in his lithograph with the snakes biting each others' tails. The strip is not divided into two pieces, as you might expect; rather, it remains a single strip. Stretch out that strip and you will see that it has become thinner and longer, but it is still one strip.

Now instead of cutting the original strip in the middle of its width, try cutting all along it at one-third of its width. When the scissors have finished their job, you will see that you have created two interlinked strips, one longer and one shorter.

It is easy to see why the Möbius strip continues to fascinate mathematicians and artists, as well as people who are just plain curious. It has been featured on stamps, sculptures, and even on commercial logos. A Portuguese bank, Banco Totta & Açores, which was formed from the merger of two banks, selected a Möbius strip as its symbol to highlight the unity of the two original financial institutions: they were two sides that became one. But the best and most imaginative portrayals are still Escher's, an artist with a profoundly intuitive geometrical knowledge of the objects he designed.

PICASSO, EINSTEIN AND THE FOURTH DIMENSION

There is an amusing story I have been told about Picasso. When he was already in his sixties, and one of the most famous artists in the world, a very wealthy elderly lady asked him to paint her portrait. The painter did not show any interest whatsoever, but the woman insisted, offering to pay him whatever he wanted. Picasso, fed up with her entreaties, dashed off half a dozen lines on a piece of paper and handed it to the lady. "That will be ten thousand dollars", he snapped. "Ten thousand?" she asked, astounded. "But you didn't even take a minute to draw that!" Picasso is said to have retorted: "A minute! You are completely wrong. It took me 60 years."

It really doesn't matter, though, whether this story is true or false. The really interesting thing is that it reminds us that an artist continues to learn throughout this life, so it is very difficult to know the precise origin of the ideas, the techniques and the necessary context for any work of art. The best we can do is to look for the indirect roots of his inspiration by studying his way of life and the intellectual climate of the era.

Years ago art historians considered that cubism, an avant-garde art movement that flourished between 1907 and 1919 and was clearly influenced by geometrical forms, could have been inspired by the revolution in mathematics and physics that took place at the beginning of the 20th century. An extremely detailed discussion of this question can be found in the classical study by Linda D. Henderson in *The Fourth Dimension and Non-Euclidian Geometry in Modern Art* (Princeton, 1983) as well as in *Einstein, Picasso,* by Arthur I. Miller (Basic Books, 2001). In this latter

N. Crato, *Figuring It Out*, DOI 10.1007/978-3-642-04833-3_30,

book, the author argues that mathematics and physics played a decisive role in the development of cubism.

Steve Martin, the author of the play *Picasso at the Lapin Agile,* seems to have read this latter book. His play, set in 1904, skillfully explores the intellectual environment at the start of the 20th century. He does this by means of dialogues between two men who never knew one another, but who both experienced this era intensely and had the feeling that they were on the verge of transforming the world: Albert Einstein (1879–1955) and Pablo Picasso (1881–1973). At the age of 25, Einstein was 1 year away from publishing his famous 1905 studies, including the article in which he set forth the theory of relativity. And Picasso, who was 2 years younger, would paint *Les demoiselles d'Avignon* just 3 years later.

Les demoiselles, one of Picasso's most revolutionary paintings, was completed in 1907. The picture, which measures approximately 8' by 7' 8", is now on display in the Museum of Modern Art in New York City (the famous MoMA), and it is one of the star pieces in the museum's collection. It marks one of the decisive moments in the history of 20th century art, and it is usually considered to have given rise to the birth of cubism.

From Picasso's sketchbooks and various other sources, it is known that the artist permitted himself long periods of reflection while preparing to paint this picture. His sketches include innumerable drawings, one of which, surprisingly, is a projection of a solid object in four dimensions. All the sketches reveal continual attempts to simplify the elements of the human form until they resemble simple geometrical figures. Later sketches are studies from different perspectives as well as that striking element of cubism employed by both Pablo Picasso and Georges Braque (1882–1963) in their subsequent work: the juxtaposition of different perspectives on the same canvas, revealing distinct points of view of the same object.

To understand how these artistic developments could have originated in the scientific preoccupations of the era, you only have to read the writings of the French mathematician Henri Poincaré (1854–1912), who is considered to have paved the way for the theory of relativity.

In his *Science et hypothèse*, a work of philosophical reflection and scientific diffusion that was published in 1902, Poincaré explains how a four-dimensional world can be represented, starting with an analogy of a projection of a two-dimensional picture on to our retina. We know that objects are three-dimensional, he says, because we perceive them sequentially from different perspectives, and we have become accustomed to representing them in two dimensions. So, he continues, a four-dimensional figure can also be represented in two dimensions. In addition, we can select from a variety of perspectives from various points of view, resulting in this sequence of visual perspectives that correspond to different projections on two dimensions from different points in four-dimensional space.

It is known that Einstein read a German translation of Poincaré's 1904 book, and that what he read had an enormous influence on his reflections about the physical world, as well as about four-dimensional space-time. It is unlikely that Picasso would have read Poincaré, but we do know that one of those who influenced his circle of friends was Maurice Princet, an actuary with an extensive knowledge of mathematics, and that Princet, who was later dubbed the "mathematician of cubism",[1] spoke to Picasso often about the fourth dimension, about non-Euclidean geometry, and about other scientific ideas that fascinated Picasso and the members of his circle.

Les demoiselles illustrates the solution Picasso found for the problem of three-dimensional representation: the simultaneous (and not sequential, like the representations that Poincaré talked about) painting of different perspectives of the same object.

It is unlikely that the artist was moved to create cubism by direct mathematical inspiration. But the similarity between Picasso's preoccupations with geometrical concepts and those of Poincaré and Einstein with space-time are too clear-cut to be a mere coincidence.

---

[1]   A. I. Miller, *Einstein, Picasso*, Basic Books, 2001, p. 100

POLLOCK'S FRACTALS

Jackson Pollock (1912–1956) is well known for his gigantic pictures that combine colored lines, splashes of paint, extensive spirals and rhythmical tracks. But he is just as well known for the controversy his art has generated. Some people have asserted that a monkey could paint more interesting pictures than Pollock's, or have commented that it is impossible to tell the difference between his pictures and completely random scrawls. How could this man have consciously created such strange, chaotic pictures?

In 1950, a photographer from New York named Hans Namuth succeeded in obtaining Pollock's permission to photograph him in action. When the appointed day arrived, Namuth turned up at the painter's studio, a barn adjacent to a farmhouse in Long Island, New York, to which Pollock had moved in 1945 after having lived in New York City for 5 years.

When Namuth arrived at the painter's farmhouse, Pollock informed him that there would be nothing to photograph that day after all, as he had finished putting the final touches to one painting and he wasn't ready to start another one yet. The two men headed for the barn, where the still damp painting was lying on the floor. They paused to admire it. Jackson Pollock began walking around the painting, and seemed obviously unsettled. Namuth couldn't imagine why. Was something missing or was there something out of place in that mishmash of apparently random lines? Suddenly the painter came to a halt, shook himself, and went off to look for a bucket of paint. He looked closely at the painting again and then started to splash paint on it. Namuth began to take photos.

N. Crato, *Figuring It Out*, DOI 10.1007/978-3-642-04833-3_31,

Some days later, the photographer showed Pollock the images he had taken. Pollock and his wife, the artist Lee Krasner, liked the photos so much that they gave Namuth carte blanche to continue to take more pictures. Namuth then spent numerous long days and evenings with the painter, taking a great many photographs. In the end, convinced that still pictures would not convey the complexity of Pollock's working methods, he produced a short video documentary. This film has been shown countless times at exhibitions, in studios and on television, and is treasured as an invaluable documentation of the artist's working day. It illuminates his technique, showing how Pollock painted a picture layer by layer, beginning with coarse applications of a background color. Next he created finer lines by means of long arm movements, allowing rivers and splashes of paint to fall. Finally, making shorter movements, he threw fine lines and small splotches on to the canvas. Namuth's film documented a very complex and not at all arbitrary technique for creating a painting.

Recently the Australian physicist Richard Taylor decided to use modern mathematical tools to analyze Pollock's technique. Taylor, who had studied art in his younger days, suspected that the visual appeal of Pollock's paintings had something to do with their similarity to images from nature, formed by chaotic processes that create fractals. He imagined that this similarity resulted from various idiosyncrasies in Pollock's technique.

In contrast to most painters, who apply paint to a canvas positioned on an easel, Pollock used a horizontal surface and let the force of gravity act on the paint. Once again in contrast to other artists, Pollock did not use brush strokes to produce fine controlled lines, but he instead let paint drop onto the canvas, or, at times, even threw it down. Taylor argued that this process is very similar to what occurs in nature, where outlines, shapes, and vegetation are all sprinkled throughout the landscape.

In order to achieve a better understanding of this technique, Richard Taylor constructed a device to drop paint in a rhythmical fashion. This device began as a simple pendulum with a nozzle. As the pendulum oscillated, it dropped paint onto a canvas on the ground. The result is a canvas covered by relatively simple lines.

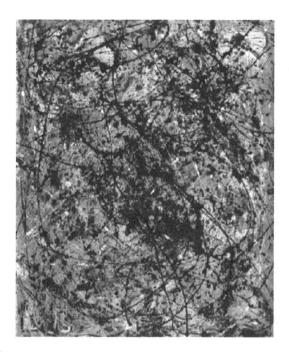

Reflection of the Big Dipper, created in 1947 by the paint-throwing method, already shows a fractal dimension close to 1.45

The image on the left is of the paint tracks left by a normal pendulum, launched time after time. The center image shows the tracks of a chaotic pendulum. On the right is a detail of "Number 14", a picture painted in 1948. The similarity between the tracks of the chaotic pendulum and Pollock's lines is striking

Then the Australian physicist decided to introduce a chaotic pendulum to drive the nozzle. Chaotic pendulums can be seen in several science museums. They are usually formed by two coupled pendular systems, which move in a complex and seemingly uncontrolled pattern, with irregularities that are impossible to predict even though they are, in fact, mostly controlled by known deterministic physical processes. Sometimes these pendulums slowly achieve equilibrium, only to then start oscillating rapidly immediately afterwards. They suddenly appear to interrupt their motion, to the surprise of an observer. Such pendulums are called "chaotic" because small modifications of the initial conditions produce radically different motion patterns after a period of time, which is why their future positions cannot be precisely predicted: As it is impossible to provide an absolutely precise characterization of the initial position and the forces acting upon the pendulum, it also becomes impossible to predict its motion over a longer period, even if there are no random elements in the system.

Pollock's 1937 painting *The Flame* does not show any clear fractal dimension yet

This is a completely different situation than we have with normal pendulums, like the ones used in grandfather clocks. In a typical pendulum, no matter what the initial starting point is, we always have a very precise idea of how it will move for quite some time after that initial point. That is why these pendulums are used to measure time.

In order to endow the pendulum with a persistently chaotic motion, Taylor constructed an electromagnetic system that pushed the pendulum periodically, but was not synchronized with the free motion of the pendulum. The paint tracks obtained as a result of this motion show great similarity to the paint tracks in Pollock's pictures, as you can see in the illustration.

The most interesting aspect is that the tracks generated by the chaotic pendulum show fractal dimensions, in contrast to those created by the simple pendulum. It is not easy to clearly understand the significance of this, as the mathematical concepts that make up the rigorous definition of fractals are complex and involve the differentiation of the so-called topological and Hausdorff-Besicovitch dimensions. But there is one geometrical property of fractal objects that is easy to comprehend: in these objects the patterns are repeated in similar form at different scales. If we look at a leaf, for example, we see veins that bifurcate into even finer veins. If we look at these very fine veins through a magnifying glass, we note that they subdivide once again into still finer veins. If we use a microscope, we can discern the same pattern. In other words, the vein structure of a leaf displays fractal characteristics.

If a leaf had only a single straight vein, it would not be classified as fractal: we can assign such a system to dimension 1. But the veins of a leaf subdivide and multiply over its entire surface. If these veins filled the leaf completely, we would have dimension 2, as they would cover the surface. But instead, what we have seen is something intermediate: as the amplification at which we observe the leaf increases, new veins previously invisible to the eye make their appearance, forming a network that nearly, but not completely, covers the entire surface. So this system of veins has a fractal dimension with a number between 1 and 2. The simpler it is, the closer it is to 1; the denser the network of veins revealed by amplification, the closer it is to 2. In statistically fractal objects, such as those encountered in nature, it is not exactly the same patterns that become visible when we change the scale, but they do display statistically similar properties.

To measure the fractal dimension of objects that live on a plane, we can divide this plane into ever-smaller squares, and then check how the patterns are repeated as the scale is changed. This was exactly the method used by Taylor and his colleagues to obtain very precise estimates of the fractal dimension of Pollock's paintings. His conclusions were very clear: the artist created paintings with a markedly fractal dimension, and as his technique advanced, he created ever more complex paintings, with a higher fractal dimension.

By 1943 Pollock's paintings had a modest fractal dimension, a little above 1. As an example, let us take *The Flame,* painted in 1937, which does not possess any marked fractality. After this period, when he had created and perfected his paint-throwing technique, he produced paintings with much clearer fractal characteristics. This applies to *Reflection of the Big Dipper,* painted in 1947, which has a fractal dimension of about 1.45, a value approaching the estimated dimension for natural structures, such as the coastline of the British Isles.

Jackson Pollock successively increased the complexity of his paintings. *Blue Poles,* a work completed in 1952, attains about 1.72, the highest fractal dimension of any of the paintings studied by Taylor. It seems that the painter was exploring the limits of what the human eye considers aesthetically pleasing. Can it be that this limit is set by nature, but then revealed via the language of mathematics?

VORONOI DIAGRAMS

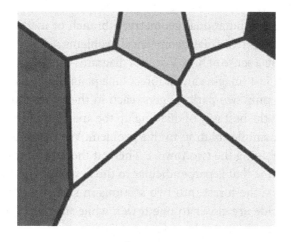

Leonel Moura, 2004

There are mathematical concepts that emerge gradually, springing from within a wide range of contexts that are apparently unconnected and of limited interest. And then, all of a sudden, these same concepts begin to attract the attention of specialists, give rise to numerous studies and applications, and end up contributing to the development of new fields of study. Voronoi diagrams are an example of this type of concept. They were first described systematically in a 1908 article by the mathematician Georgi Voronoi, but their roots can be traced all the way back to ideas first presented in 1644 by René Descartes on the distribution of the planets in the solar system, as well as to work undertaken in the middle of the 19th century by the German mathematician G. L. Dirichlet.

Georgi Feodosevich Voronoi was born in 1869 in Zhuravki, (a town that was then in the Russian Empire, but is now in the Ukraine). Voronoi studied at the University of St. Petersburg, where he was awarded a Ph.D. in 1897. He died in Warsaw at the age of 40 in 1908, after spending his short life working as a university professor, mainly in the fields of number theory and computational geometry problems. In the last year of his life he published a study exploring and then defining a division of space that came to be called a Voronoi diagram. His concept was picked up and reapplied in 1911 for meteorological studies, and in 1927 once again used in the study of crystallography. Currently the Voronoi diagram is used in a wide variety of areas, and is studied in the field of computational geometry, a branch of mathematics that applies algorithms to resolve geometrical problems.

To give you a sense of how Voronoi diagrams can be applied in real-life situations, just imagine a vast forest on a plain that is monitored for forest fires by only two park rangers, each in their own individual fire tower. What is the best way of dividing up the area they are monitoring?

There is a simple solution to this problem. We can draw a straight line segment joining the two towers. Then, at the midpoint of this segment, draw a line that is perpendicular to the first one. This second line serves to divide the forest into two sections in such a way that all the trees on one side are closer to one tower, while the trees on the other side are closer to the second tower.

If there were three fire towers in the forest that were not all positioned on one straight line, our problem would become slightly more difficult, although its solution remains obvious. In this case, we would have to divide the forest into three sections bounded by half lines. It is not too challenging to figure out how to calculate the position of these lines. However, if there were ten towers distributed randomly throughout the forest, it would become far more difficult to work out a solution. Generally speaking, this situation would require that the forest be divided into sections shaped as (closed) polygons and non-limited figures, a type of polygon with one side missing.

This example perfectly illustrates what a Voronoi diagram is: there is a plane and a set of points located on it. That plane is divided into cells, each of which is defined by grouping together all the locations that are closer to one point than to any other point. The boundaries between these cells form the Voronoi diagram generated by this given set of points.

These diagrams clearly have applications in many situations. Our fire tower example can be modified, but it will still generate exactly the same figures. Imagine, for instance, that a fire was started at each tower at a specified moment, and that the fires all advanced at the same speed in every direction. After a certain time, the fronts of the fires, which began as circumferences centered on each tower, would meet. There would be trees that were burnt back and front by different fires coming from different directions. If we then marked the positions of these trees, we would see that they were equidistant from two or more of the sources at the fire towers. If we then drew lines joining the positions of these trees, we would end up with an exact duplication of the Voronoi diagram from our previous example.

This example leads us directly to some applications of the Voronoi diagram for crystallography. Imagine a set of dispersed points around which a material crystallizes at a constant growth rate. What shape would the edges of these crystals have when each of them had finished growing, inherently limited by their collisions with the other crystals growing beside them? The answer is a Voronoi diagram.

There are also interesting applications of Voronoi diagrams for robotics and optimum control. Imagine a set of obstacles placed at points dispersed on a plane a robot must cross. If we wanted to select a trajectory on which the robot stays as far away as possible from each obstacle (the points), such a trajectory would be part of a Voronoi diagram generated by the obstacles. If we imagine that the vertices of the resulting Voronoi diagram are locations, then we can see the similarity with the famous problem of the traveling salesman who has to find the shortest route enabling him to visit each of the stops on his list.

Voronoi diagrams can have countless other applications as well. They can even become the subject of artistic endeavors, such as those of the Portuguese artist Leonel Moura, who painted a series of pictures he called *Algorithms on Canvas*. Moura used a computer to generate Voronoi diagrams, and then he colored the spaces in between. Does the beauty of these figures derive from their conceptual simplicity, from the equilibrium we see in them? At times it is not easy to discern the point where the rigor of mathematics ceases and the hand of the artist takes over. Maybe that is the artist's secret, and the mathematician's.

THE PLATONIC SOLIDS

Have you ever looked closely at a cube? It is one of the commonest solids. In nature it appears in crystals; in our homes in furniture design; in casinos, cubes can be seen in the dice rolled on betting tables.

Mathematicians classify the cube as a polyhedron, a closed three-dimensional geometric figure limited by planes. The faces of the cube are squares, a common feature of polygons (plane figures bounded by edges that are segments of straight lines).

Taking a closer look at the cube, we can see that it has six faces, all of them equal, and also 12 edges, all of the same length. The 8 vertices of a cube also have one characteristic in common: three edges are joined at each vertex. In other words, a cube has equal faces, equal edges, and vertices that all join the same number of edges.

There are many solids that are not as symmetrical as a cube. For example, the pyramids in Egypt have sides formed by four triangles, but their bases are square. But we can find other solids with the same type of regularity as a cube; take the tetrahedron for instance, a solid formed by four equilateral triangles.

The Ancient Greeks took a great interest in the regularity of geometrical shapes. Some people even claim that it was their interest in abstract patterns that launched the Greek intellectual miracle, which eventually enabled western civilization to make such extraordinary progress. The Greeks discovered a total of five solids with a degree of regularity similar to that of the cube. Plato (about 428–348 BC) discusses them at length in his dialogue *Timaeus*. That is why these five geometrical figures have come to be known as the *Platonic solids*.

N. Crato, *Figuring It Out*, DOI 10.1007/978-3-642-04833-3_33,
© Springer-Verlag Berlin Heidelberg 2010

The most curious aspect of this is that there are only five solids of this type. If we restrict ourselves to polyhedrons ("figures with sides and angles that are equal and equal to each other" as defined by Euclid about 300 BC), these five are the tetrahedron, the cube, the octahedron, the dodecahedron and the icosahedron. Euclid knew of a proof of this fact, which he included in his *Elements*. But it is likely that it was known even long before Euclid's time that it is impossible to construct any other solids with these precise characteristics.

For followers of Plato and of Pythagoras, these polyhedrons provided proof of the intrinsic harmony of the world. These devotees of Plato and Pythagoras studied many of the aspects of polyhedrons and discerned a series of curious relationships: while the cube has 6 faces and 8 vertices, the octahedron has 6 vertices and 8 faces. While the dodecahedron has 12 faces and 20 vertices, the icosahedron has 12 vertices and 20 faces. This means that we can assign these four solids into two sets of pairs, the cube with the octahedron and the dodecahedron with the icosahedron. Such pairs are called duals. The tetrahedron, however, remains alone, with its 4 faces and 4 vertices it is its own dual.

Some of these solids are rigid. If we construct the edges of a cube using strong rods and flexible adhesive, we will see that it can be destabilized by twisting or bending it. This does not happen with the octahedron, the dual of the cube. Once we have joined the rods that form its sides, the octahedron cannot be manipulated. You can only twist it by first dislodging its rods altogether. This also occurs with the other pair of dual solids: the icosahedron is rigid, but the dodecahedron can be bent. Our single, the tetrahedron, is rigid.

The Greeks were awed by these relationships, and believed that they revealed hidden secrets of the universe, even of the physical and contingent universe. Plato contended that earth is represented by the cube, fire by the tetrahedron, air by the octahedron and water by the icosahedron. The outsider was the dodecahedron, which represented the cosmos. We may today smile at these suppositions, but the truth is they still reveal a profound knowledge of three-dimensional geometry.

Another characteristic of the Platonic solids is their geometrical symmetry with respect to a center. Any one of these solids can be inscribed in a sphere, with all its vertices touching the surface of the sphere. This means that all the vertices are at the same distance from the central point of the solid. But, in addition to an *outer sphere* that circumscribes the polyhedron, we can draw an inscribed *inner sphere* that touches the central point of each of the faces of the solid.

The Greeks were not the only ones to become fascinated by the symmetry of the Platonic solids. Almost 2000 years after Plato, Johannes Kepler (1571–1630), one of the greatest scientists ever, thought that he had found the secret of the world's harmony in the perfection of geometry. This German astronomer constructed a model of the solar system using spheres circumscribed in Platonic solids. Everything seemed to be in perfect harmony. The sun was at the center, itself forming a sphere. Around it was another sphere inscribed inside an imaginary octahedron, touching the centers of its faces. Mercury, the nearest planet to the sun, moved within this sphere. Another sphere, touching the vertices of the polyhedron, was inscribed around this octahedron. The planet Venus moved within this second sphere, which was inscribed inside an

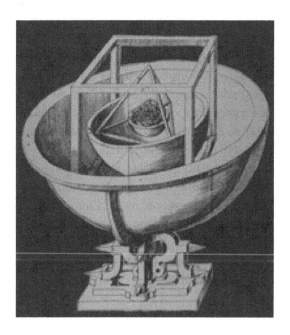

icosahedron, touching the centers of its faces. The icosahedron was, in turn, circumscribed around another sphere, where the Earth orbited and so on. As only six planets had thus far been discovered, the six spheres separated by their five solids seemed to reveal the hidden harmony of the universe.

However, Kepler loved the truth, and never settled for mere assumptions. Although he was convinced for some time that he had indeed found the secret of the cosmos, he came to accept the factual evidence. After spending many years studying the observations collected by his master, Tycho Brahe (1546–1601), he perceived that his model did not correspond to the facts. The relative distances between the planets did not agree with the requirements imposed by his theory of the spheres. And later Kepler discovered that the planetary orbits were not circular, but elliptical, a fact that he revealed to the world in what came to be known as Kepler's First Law.

A seventh planet in the solar system was discovered in 1781 by the astronomer William Herschel (1738–1822), an event that once again demolished the model of the solar system based on Platonic solids. But not everyone learned from the humility shown by Kepler. Two hundred years later, the philosopher William Georg Friedrich Hegel (1770–1831) remained convinced that the solar system consisted of only seven planets, as seven was a divine number and the universe was the material revelation of spiritual harmony. Hegel was still alive when the first minor planet, the asteroid Ceres, was discovered in 1801 by the Italian astronomer Giuseppe Piazzi (1746–1826). At that time this eighth solar satellite was considered to be a planet, and its discovery considerably tarnished the scientific reputation of Hegel, a philosopher who delighted in speaking of mathematical and physical concepts that he did not understand (a defect that may still be noted in certain modern-day intellectual impostors). In 1846 the planet Neptune was discovered; and then Pluto in 1930. Science may have learned to study the geometry and symmetry of solids, but it also learned to be wary of speculative theories of the Platonic harmony of the universe.

PYTHAGOREAN MOSQUITOES

Before people have regular exposure to other cultures, there is a natural tendency to assume that everything that happens in the world mirrors our own experiences. We might think, for example, that everyone everywhere else eats a fried breakfast, or that there is no city on earth more hospitable than our own. But then one day we encounter sushi, or read the work of writer Machado de Assis, and begin to see that there are many ways of living life that are different from our own way.

The discovery of such a variety of cultures, criteria, and creeds was a very positive development in the 20th century. Many western thinkers began to see that embracing cultural differences was both necessary and beneficial. Artists and musicians also agreed, thinking up new ways to paint and creating musical compositions that, in the old days, would have been dismissed as noise. We came to see that aesthetic criteria can vary according to a person's culture and education.

One school of thought whose influence spread from literary criticism to sociology and philosophy took this perspective to an extreme: it maintained that there is nothing in art, ethics, politics, and even in science that is absolute, that does not depend for its relevance on the culture through which it is being viewed. In other words, everything would be a "social construction", as they began to call it.

Can this be right? Does this mean science is not an objective discipline? That there is no confrontation with external reality? And even in art? Excrement repels us and beautiful sunsets delight us. Is this simply due to a social construction? Consider the passion Pythagoras felt when

N. Crato, *Figuring It Out*, DOI 10.1007/978-3-642-04833-3_34,
© Springer-Verlag Berlin Heidelberg 2010

he considered the interplay of numbers in musical harmony. Is this just another social construction?

Legend has it that Pythagoras studied the proportions between the various sizes of blacksmith's hammers and ascertained that some produced harmonies while others created dissonance. Pythagoreans were later amazed when it was found that this discovery could be applied more generally. For example, they deduced that the harmonic proportions in string instruments correspond to strings with length proportions that can be expressed as ratios of whole numbers.

Today when we speak of sound frequency, we measure it in cycles per second or Hertz (Hz). When the frequency of a sound is doubled, for instance, it becomes higher, going up one octave in the musical scale. When a frequency is multiplied by 3/2 it goes up by a fifth, and so on. The resulting chords are pleasing to the ear. The amazing thing about this is that nature apparently thinks so too.

Four scientists from Cornell recently published an article in *Science* about their investigation into the mating songs of the yellow fever mosquito (*Aedes aegypti*).[1] They did not actually hear "amore mio" in insect language, but they did manage to establish that the mating ritual of these mosquitoes is preceded by a tuning of frequencies in a common harmony.

Here's how it works: the males start by using their wings to produce sounds at a frequency of 600 Hz – six hundred cycles per second –, or about the same as G in our scale. But when they begin to really pursue the females, they double the frequency, producing a note similar to G one octave higher. The females, who normally produce sounds at a frequency of 400 Hz (close to C on a piano), respond to the males by tripling the frequency of their own wing beats so that they tune in with the males at 1200 Hz, which is the smallest common multiple of the two primary frequencies. The resulting note, G, is the mosquito's music of love. Pythagoras himself could not have done better.

---

[1] L. J. Cator, B. J. Arthur, L. C. Harrington, and R. R. Hoy, Harmonic convergence in the love songs of the dengue vector mosquito, *Science* **323**, 1077–1079, 20 February 2009.

THE MOST BEAUTIFUL OF ALL

How can we find beauty in an equation? Readers will certainly have divided opinions about this. Some people will assume the question is ironic: what possible beauty could there be in those incomprehensible squiggles that filled our schoolbooks? But in the view of others who carried on with math after leaving school, and came to enjoy this subject, even making it their life's work, the simplicity and elegance of certain equations make them beautiful. Quite beautiful, in fact, though in some cases it is difficult to explain the reasons for their beauty. One is surely the strange condensation of reality they conveyed, reality that may be geometrical, physical, biological, or purely ideal. And then there is their flexibility, their applicability to infinite numbers of unexpected situations, as well as their graphic representation.

In mathematics, nobody is surprised by the existence of equations; the condensation of relationships by means of symbols seems to define science itself. But they also illuminate the enormous explanatory and predictive power of science.

For example, the brilliance of physics is that it uses equations to interpret very general laws, and permits forecasts to be made that reveal unsuspected aspects of reality. In many cases, for instance in Einstein's most celebrated equation, $E = mc^2$, the quest to explore mathematical relationships led scientists to formulate new questions and conclusions. This seems to signify that equations contain more than their creators put into them, that perhaps they are "wiser than their discoverers",[1] as the

---

[1] E.T. Bell, *Men of Mathematics,* New York, Simon & Schuster, 1937, p. 16.

N. Crato, *Figuring It Out*, DOI 10.1007/978-3-642-04833-3_35,
© Springer-Verlag Berlin Heidelberg 2010

German physicist Heinrich Hertz once put it. Graham Farmelo, the editor of *It Must Be Beautiful: Great Equations in Modern Science*, observed in this book that an equation is an expression of perfect equilibrium, and what makes it beautiful is its synthesis of truth without a single wasted symbol. The various essays that comprise the book discuss the extent to which aesthetics is a factor in the work of mathematicians and scientists. For example, Einstein declared that "the only physical theories we are willing to accept are the beautiful ones".[2] Another physicist, Paul Dirac (1902–1984), went farther, and used a bit of a hyperbole when he claimed that "it is more important to have beauty in one's equations than to have them fit experiment".[3]

In the aftermath of Graham Farmelo's book, Robert P. Crease, a columnist for the journal *Physics World*, initiated an opinion poll among his readers, asking them to list the equations they thought were most significant in the whole history of science, taking into consideration their reach, depth and aesthetic appeal. Their responses are revealing.

His readers tended to favor equations that are at once important and also notably simple. The most popular was a purely mathematical equation formulated by the Swiss mathematician Leonhard Euler (1707–1783). Following closely behind came a series of equations created by the Scottish physicist James Clerk Maxwell (1831–1879) that describe the behavior of an electromagnetic field.

Mathematicians were not surprised to learn that Euler's equation had come out on top. Using only seven symbols, it includes three basic operations and states the relationship between the five most important numbers in mathematics. Its beauty derives from its simultaneous simplicity, and its profundity. Have a look:

$$e^{i\pi} + 1 = 0$$

[2] Graham Farmelo, *It Must Be Beautiful*, London, Granta, 2002, p. xiii.
[3] Paul Dirac, The Evolution of the Physicist's Picture of Nature', *Scientific American*, May 1963, 208, p. 47.

Euler's equation contains the two most important integers, 1 (which is the unit from which it is possible to find all the integers and rational numbers by using only four elementary operations) and 0 (which unleashed a revolution in mathematics as it constitutes a higher abstraction). Then comes π, the quotient of the circumference and its diameter, an irrational number that, even today, has not yet revealed all its mysteries, as well as two other numbers that are ubiquitous in advanced mathematics.

One of these numbers is the "imaginary unit", the square root of −1, a number that was created in order to resolve algebraic problems and has since been shown to have a right to exist. "Complex numbers", which encompass real numbers, can be written as sums of real and imaginary numbers. Complex number have tremendous interest for mathematics, and have direct applications in physics and in such practical areas as electronic engineering.

The other fundamental number included in Euler's equation is the base of natural logarithms, the number $e = 2.71828\ldots$ This is such an important number that it plays a role in such disparate topics as compound interest, population growth, and radioactive decay, as well as in the spirals found in flowers and galaxies. In pure mathematics, it can be found in the definition of angles, and as the limit of important series and sequences; it also appears in derivation, integration and other analytical operations. As far as we can judge, it came to light for the first time in 1618, seemingly emerging from a simple problem in compound interest computation. It is truly astonishing that the very same number appears in such diverse fields and has so many practical applications. Maybe it is even more amazing that it can equal the unit after being raised to the imaginary unit multiplied by π. Euler's equation is a celebration of the unity of mathematics, and of the power of science. It is not at all surprising that many people consider it the most beautiful equation that has ever been written.

MATHEMATICAL OBJECTS

THE POWER OF MATH

"How can it be" wondered Einstein, "that mathematics, being after all a product of human thought which is independent of experience, is so admirably appropriate to the objects of reality? Is human reason, then, without experience, merely by taking thought, able to fathom the properties of real things?"[1]

This question is neither naïve nor ingenuous. As the great physicist observed in a 1921 lecture on "Geometry and Experience", "At this point an enigma presents itself, which in all ages has agitated inquiring minds."[2] The response to this enigma has divided mathematicians and philosophers. While some regard the applicability of mathematics as the natural product of its roots in our experience, others consider that its success corresponds necessarily to the real world with the logical premises on which mathematics is based. Einstein relativized these matters in a point of view that is supported by many mathematicians, scientists and philosophers. "As far as the propositions of mathematics refer to reality, they are not certain; and as far as they are certain, they do not refer to reality"[3] he said.

According to Einstein, modern mathematics, founded on a logical-formal deduction based on axioms, has succeeded in separating its logical-formal aspect from its objective and intuitive content. The

---

[1] The English translation of Einstein's 1921 lecture is available online at http://pascal.iseg.utl. pt/~ncrato/Math/Einstein.htm

[2] Ibid.

[3] Ibid.

N. Crato, *Figuring It Out*, DOI 10.1007/978-3-642-04833-3_36,
© Springer-Verlag Berlin Heidelberg 2010

correspondence of mathematical conclusions to physical reality is merely approximate, and is derived from the possible approximation of the axioms to fundamental natural laws.

Ian Stewart, a prolific English mathematician who has also specialized, very successfully, in the field of science communication, does have an answer to these questions. In his book *Nature's Numbers*, published by Basic Books, he recognizes that there are various theories that explain the usefulness of mathematics, "ranging from the structure of the human mind to the idea that the universe is somehow built from little bits of mathematics".[4] But his response is "quite simple: mathematics is the science of patterns, and nature exploits just about every pattern that there is".[5]

In his book Stewart attempts to show how the mathematical investigation of patterns can explain many phenomena found in nature. For instance, he brings up the old question regarding the spiral pattern of the shells of snails, whelks and similar creatures that was posed at the beginning of the century by the Scottish zoologist D'Arcy Thompson in his book *On Growth and Form*, a classical work written in 1917 and still in print today.

The spiral structure of the shells of these animals may be explained by means of geometry. If we accept that the shell develops as the animal grows, that it always develops in a similar way, and that the width of the developing shell tube depends on the size of the animal at any given moment, then it is natural that the rings that are produced take the form of the spirals that really occur in nature. The relationship between the width of the rings and the dimensions of the animal at the time it is creating them gives rise to various types of spirals that can be described via well-established mathematical equations.

This example permits us to return to Einstein's argument. The perfect geometrical spirals resulting from mathematical functions are not found in the natural world. The ones found in nature are imperfect. As is sometimes said, perfect points, straight lines and triangles do not

---

[4]   I. Stewart, *Nature's Numbers*, Basic Books, New York, NY, 1995, pages 18
[5]   Ibid.

exist outside our minds, but analyzing these perfect objects using precise logic helps us to draw conclusions with respect to the imperfect and approximate points, straight lines and triangles that do exist in nature.

Another interesting example of a mathematical pattern found in the real world is the arrangement of petals and florets (the small rudimentary flowers that are found, easily visible, in the center of some flowers such as sunflowers). In some species these florets are distributed in groups of spirals that curl in different directions and intercept each other. Often the number of elements that curl in one direction is 34, while the number of elements curling in the opposite direction is 55. Other cases involve pairs of spirals consisting of 55 and 89, or even 89 and 144 elements.

This may seem to be just another curious fact, but mathematicians look at these numbers and recognize that they are consecutive terms in the sequence 1, 1, 2, 3, 5, ..., 34, 55, 89, 144, 233, ... As we have seen before, this is a numerical sequence constructed in 1202 by Leonardo of Pisa (1170–1250), also known as Fibonacci, in a discussion of a problem involving the population growth of rabbits. In this sequence all the terms after the first one are obtained by adding together the two previous terms $(2 = 1 + 1, 3 = 1 + 2, 5 = 3 + 2,$ etc.).

Why do these numbers, created in response to such a different problem, recur in the elements of flowers? Biologists may be tempted to say that they are numbers that are found in the genes of the plants, but mathematicians search for other reasons. The genes determine how a being develops, but it develops in the context of a physical and geometrical world where restrictions also exist. Mathematicians have succeeded in demonstrating that elements that develop around a central point in such a way that they occupy the surface area in the most compact manner possible achieve this by means of a precise angle of divergence, the "golden angle" (approx. 137.5°). Well, circular elements that develop in such a way that they are always separated by this angle tend to form spirals in which the Fibonacci numbers occur. So it is not surprising that this angle and these numbers are found with great regularity in flowers, both in their petals and their florets.

Mathematics succeeds in explaining the geometrical and numerical regularity on the basis of very simple principles of growth that may be determined by the genes. But the living world does not need to have all the rules of mathematics inscribed in a DNA code. Such rules occur naturally, based on even simpler rules of growth. After all, these mathematical patterns are patterns that are necessary in nature. Could this explain why mathematics is "so admirably appropriate to the objects of reality"?

DOUBTS IN THE REALM OF CERTAINTY

In the early 1980s Philip Davis and Reuben Hersh wrote a bestselling book entitled *The Mathematical Experience*. The book's philosophical message is found throughout the narrative of the intellectual adventure that has given rise to modern mathematics. In another book by Hersh published in 1997, called *What is Mathematics, Really?*, the author makes his philosophical leanings clear. It was enough to re-ignite a lively debate on the fundamental aspects of this mysterious entity that is mathematics.

Everybody knows, or thinks they know, what mathematics is: a game of numbers, unknowns and relationships used to pay the bill at the supermarket, calculate taxes, design bridges, determine the trajectory of the space shuttle and reassure us about the future orbits of asteroids. But philosophers, just like mathematicians, have never agreed on the roots of the miraculous power of these numbers and equations.

Throughout the centuries, two main currents of thought have vied for supremacy: Platonism and formalism. It is difficult to reduce these currents to more general schools of philosophy, such as materialism or idealism, for it is a debate that is unique to the world of mathematics.

Platonism, sometimes also referred to as realism, asserts that mathematical entities exist outside space and time, outside the human domain, independently of our existence. Accordingly, the objective of mathematics would be to uncover the various aspects of a grandiose abstract structure that is composed of objective, unquestionable and immortal truths.

N. Crato, *Figuring It Out*, DOI 10.1007/978-3-642-04833-3_37,
© Springer-Verlag Berlin Heidelberg 2010

The Hungarian mathematician Paul Erdős (1913–1996) used to say that God had a book in which the best proofs of all the theorems were written – the perfect proofs. He considered that his task as a mathematician was to discover what was in that book, although God did not always play along. Of course Erdős was being ironic, but his comment describes the Platonic point of view.

Formalism, on the other hand, views mathematics as an axiomatic and logical construction, as a "game" of no significance. Therefore it should be possible to construct various truths depending on the presuppositions (the *axioms*) that are made. As an external reality approaches the presuppositions of a theory, so the results of that theory can be applied to that reality. But always only approximately.

Einstein was not far from this point of view. "In my opinion", the physicist said in defense of the axiomatic approach, "the answer to this question is, briefly, this: as far as the propositions of mathematics refer to reality, they are not certain; and as far as they are certain, they do not refer to reality."[1] So where does the power of mathematics come from? Two and two seem to make four, both in the mental addition derived from the rules of algebra as well as when counting and adding up real oranges.

The greatest apostle of formalism was David Hilbert (1862–1943), a German mathematician who taught in Göttingen. At the turn of the 20th century he proposed that the basic tenets of mathematics should be subjected to absolutely rigorous revision. Hilbert urged that the whole of mathematics should be based on a finite set of axioms from which all results should be derived by logical rules. "The goal of my theory", he stated, "is to establish once and for all the certitude of mathematical methods."[2]

In 1931, Kurt Gödel (1906–1978), a logician of Austrian origin who worked in Vienna and at Princeton, was able to demonstrate that such

---

[1]   See http://pascal.iseg.utl.pt/~ncrato/Math/Einstein.htm

[2]   D. Hilbert, "On the infinite", 1925 address, as translated and reproduced in P. Benacerraf and H. Putman, editors, *Philosophy of Mathematics: Selected Readings* (2nd edition), Cambridge University Press, 1983, p.184

an outcome was unattainable. Given any set of axioms, there are always mathematical results that cannot be proven using it. But Gödel and his famous theorems of incompleteness did not annihilate the formalist vision, although they did limit its dream of perfection.

Reuben Hersh quips that most "working mathematicians" vacillate between these two currents of thinking, being Platonists on weekdays and formalists on weekends. "On weekdays, when doing mathematics, he's a Platonist," writes Hersh, "he is convinced he's dealing with an objective reality whose properties he's trying to determine. On weekends, if challenged to give a philosophical account of this reality, it's easiest to pretend he doesn't believe in it."[3]

Hersh endeavors to bypass this dilemma by promoting a new theory, one of mathematics as a social construct. He supports it by invoking the theses of Imre Lakatos (1922–1974), a philosopher of Hungarian origin who worked with Karl Popper (1902–1995) but later disagreed with him, and who also worked with the mathematician George Pólya (1887–1985). Lakatos emphasized the fallible nature of mathematics, observing that it takes problems and conjectures as its starting points and then flourishes by means of the criticism and correction of theories, which are always susceptible to ambiguity and error. It is not by chance that the magnum opus of Lakatos is entitled *Proofs and Refutations*.

In his search for a new philosophy of mathematics, Hersh explores its framework, i.e., the manner in which it is constructed. It would seem to be a perfect and precise construction, and this is the image purveyed by scientific articles, books and textbooks. But this framework conceals a construction that at times vacillates, is subject to advances and reverses, uncovers errors, and leads to innumerable blind alleys. Hersh emphasizes that, in fact, many truths that were once considered absolute have since been shown to be imperfect. Many of the steps in mathematical reasoning suffered at the hands of subtle errors, discovered only at a later date.

[3] P. 39.

Nobody doubts that the construction of contemporary math can contain errors and imprecisions. But this human construction may reveal something about our universe that cannot be explained simply by the actions of society. Hersh's detractors accuse him of confusing the construction process of mathematics with the discussion of its reality and applicability. There is no doubt that there are errors and restrictions in the construction process, as there are in every activity undertaken by human beings. But it is the confrontation with reality that gradually corrects these restrictions and errors. Mathematics, like science, cannot only be a social construct. Logic, coherence and reality are its points of reference. There is indeed a reason for the unreasonable effectiveness of mathematics.

When Chance Enhances Reliability

Randomized algorithms have revolutionized the way mathematics operates. Surprisingly, making use of chance may be the fastest way to obtain the solution to a problem, and not only when it comes to math.

You will certainly have experienced something like this: you're calling a friend and suddenly the connection is lost. You call your friend back but now the line is busy, as he is also trying to call you. You hang up and wait a few seconds to see if the phone will ring, but of course it doesn't, as your friend is also waiting to see what will happen. So you call again, but once more the line is blocked... and so it goes on until you and your friend stop making the same decisions at the same time, and then finally one of you gets through. This is a typical situation involving a conflict of rational patterns, and a successful outcome is achieved only by the chance occurrence of the decisions being taken out of phase. It is difficult to find a solution for this problem, since only two contrary solutions that have to occur simultaneously (one person waits and the other calls) breaks the impasse. If there is no agreement between the two persons from the start that upsets the symmetry between their actions, then there is no optimum process for solving the problem. In the end it is chance that allows the two friends to finally reconnect.

The same thing is true for many mathematical algorithms, that is for many finite sequences of rules that lead to a solution, whether exact or approximate, of a clearly formulated problem. The deterministic algorithms used to solve many such problems have limitations that can only be overcome by chance.

N. Crato, *Figuring It Out*, DOI 10.1007/978-3-642-04833-3_38,

One example that illustrates this point is the *QuickSort* algorithm, a process used to sort a series of elements, such as a set of names that are to be listed in alphabetical order. This algorithm is very common in computer programs. It compares the elements of the list, starting with the first, by making pairwise sequential comparisons, correcting any incorrect sequences that it finds, and then proceeding to the next element on the list. If the elements in the series are randomly mixed, the algorithm is very efficient, as it can minimize the number of comparisons.

The efficiency of a sorting algorithm can be measured by the number of comparisons it makes. If we are unlucky, or the algorithm is not very efficient, we can end up having to compare each element one by one with every other one, which is the most inefficient method of all.

Surprisingly, this is what we find when the list has already been sorted and we want to sort it again or double check the order. In this situation the *QuickSort* method is the worst possible solution. The algorithm will compare the first name with each of the others and then leave it in its original position. It will then compare the second name with all the subsequent names, etc., eventually comparing every possible pair of names.

It seems difficult to find a better procedure. Beginning with the first name is the worst possible way to begin. What about starting with the last name? In that case we potentially encounter the same problem. What if we were to start in the middle? It would not work well if the original list has blocks with ordered names. . .

It seems we are facing the same dilemma as before: do we call or wait to be called? If our decision is always the same as the one our friend makes at the same time, we will never get to talk to one another.

What we need is a method of selecting an initial element that has the lowest probability of coinciding with a sorting element of the list. So this problem does not have a deterministic solution. Nothing is better than simply tossing a coin and allowing chance to determine our choice.

A modified version of this algorithm, the *Random QuickSort*, does exactly that. It starts by randomly selecting an element from the list that is to be compared with all the others, effectively dividing the list into two subgroups. Within each of the two subgroups the algorithm makes a random selection of the element that it will use as its reference point, and keeps repeating this process until the list has been completely sorted. The probability that the elements selected will systematically be the last elements in each subgroup is so small that usually the algorithm is extremely efficient.

Randomized algorithms like this that always find a solution to the problem are known as *Las Vegas type algorithms*. These algorithms have their origin in another older method called the *Monte Carlo method*. Both names refer to the famous gambling centers. Unlike Las Vegas algorithms, Monte Carlo algorithms are not guaranteed to always find a solution, but they do affect the desired degree of reduction of the probability and magnitude of an error.

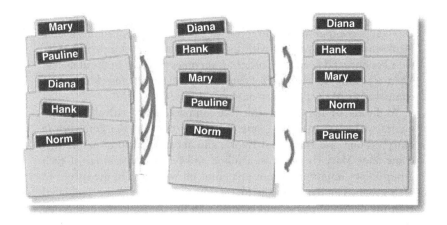

This basic idea was the brainchild of Stanislaw Ulam (1909–1984), a mathematician of Polish origin who worked in the U.S. from 1936 onwards, and was involved in some of the greatest intellectual adventures of the century. Ulam belonged to the famous Los Alamos group that included Oppenheimer, Teller, Fermi, von Neumann, Feynman and so many other celebrated physicists and mathematicians. It was this group that developed the atomic bomb and pioneered automatic computation.

As Ulam relates in his autobiography, *Adventures of a Mathematician,* "the idea for what was later called the Monte Carlo method occurred to me when I was playing solitaire during my illness. I noticed that it may be much more practical to get an idea of the probability of the successful outcome of a solitaire game [...] by laying down the cards, experimenting with the process and merely noticing what proportion comes out successful, rather than to try to compute all the combinatorial possibilities, which are an exponentially increasing number [...] This is intellectually surprising, and if not exactly humiliating, gives one a feeling of modesty about the limits of rational or traditional thinking".[1]

PUTTING THE HOUSE IN ORDER

Let us suppose that we have to sort a list of names: Mary, Pauline, Diana, Hank and Norm. We start with the first two names, comparing Mary with Pauline, and keep their order, as M comes before P in the alphabet. Then we compare Mary with Diana and put Diana before Mary, as D comes before M. And we keep on doing this.

At the end of this step, in which we compared Mary with all the other names, we have created two groups: one group of names that come before Mary (in our case Diana and Hank), and another group of names that come after Mary (in this case Pauline and Norm). The order of each group is not important. From this point on we compare the names in each group. The names in the first group do not have to be compared with the names in the second group, as all the names in the first group are before Mary and all those in the second group come after Mary. Two more comparisons (Diana and Hank in the first group, and Pauline and Norm in the second) are sufficient to sort our list completely. We have made a total of 6 comparisons to sort the list. If we had needed to compare each name with every other one we would have had to make 10 comparisons.

[1]   S.M. Ulan, *Adventures of a Mathematician,* University of California Press, Berkeley, CA, 1991, p. 196–197.

This example shows how *QuickSort* works. If the list had been longer, we would have had to create subgroups from the first two groups, and then more subgroups from the first subgroups, and so on until each of the final subgroups contained a single name, which would mean that the list had been completely sorted.

In our example the difference between 6 and 10 comparisons does not seem very substantial. But just imagine if you had a list of twenty thousand names of subscribers to a newspaper. An inefficient algorithm would have to make more than 200 million comparisons, so any time saved would be very welcome indeed.

It was 1946. Stanislaw Ulam was discussing his idea with John von Neumann (1903–1957), a mathematician of Hungarian origin who had settled in the U.S. and was also working at the Los Alamos laboratory at that time. Von Neumann, the architect of the electronic computer, immediately recognized the potential of the idea, and the two men started to apply the method to difficult calculations for the construction of the nuclear bomb. Today, innumerable simulation methods are based on the Monte Carlo method.

It is impossible to enumerate all the types of problem that are resolved today by randomized methods: calculating areas and volumes, studying engineering projects, forecasting the weather, modeling the behavior of markets, an endless number of theoretical or applied problems.

If chance often enhances reliability in scientific applications, why don't we use it in our daily lives as well? Well, now you know... the next time you lose a phone connection, why not just toss a coin?

THE DIFFICULTY OF CHANCE

At first glance nothing would seem to be as natural and easy as chance. It is orderliness and organization that seem difficult to achieve. But reality is very different. Things frequently take on lives of their own, organizing themselves and creating patterns. In a forest, for example, each type of tree tends to appear in specific areas, simply because the greatest concentration of the seeds of that tree end up in those areas. In the oceans, waves seem to move in concert, in groups, in consecutive lines, because the water masses move in coordination. In the universe, the stars appear to be grouped into galaxies, which in turn gather in groups, and then those groups cluster as superconglomerates.

This natural tendency to organize prompts us to think that pure chance must be something completely disorganized, in which there are no discernible patterns. But it is difficult to reproduce pure chance. Ask people to speak syllables haphazardly, as if inventing a new language. After just a few "words", everybody will no doubt begin to repeat the same syllables, as it is very difficult to produce really random sounds.

In one famous experiment psychologists asked people to write sequences of zeros and ones by imagining they were tossing a coin and writing 0 every time heads appeared and 1 every time tails appeared. Then they asked the same participants to write another sequence of ones and zeros, but this time after tossing a coin and each time recording the results. It turns out that if the sequences are sufficiently long, let's say, containing 20 or more digits, statistical analysis can normally find out if the result is a purely random sequence, or if the outcome was imagined. You can try this by writing a sequence of zeros and ones on a piece of

N. Crato, *Figuring It Out*, DOI 10.1007/978-3-642-04833-3_39,
© Springer-Verlag Berlin Heidelberg 2010

paper. If you want the experiment to be really conclusive, be patient and write out about 250 digits and then toss a coin 250 times and write down each of the results.

Various methods can be used to assess whether the sequences are random. For example, we can count the number of zeros, which ought to be approximately half of the total. That is, if we calculate the mean value of the digits, we should obtain a result of 0.5, or something very close to that. When a coin is tossed, it has been shown that this is what happens, i.e., "heads" and "tails" are obtained in very similar numbers. In the 18th century the French naturalist Georges Louis Leclerc (1707–1788), known to mathematicians as Count de Buffon, decided to try this experiment. He, or perhaps one of his servants, tossed a coin 4040 times, and obtained heads 2084 times, which is an average of 0.5069. Later, in the 20th century the English statistician Karl Pearson (1857–1936) repeated this experiment, tossing a coin 24,000 times, and obtaining heads 12,012 times, or an average of 0.5005. Then during the Second World War, an English prisoner of war occupied his time in the same way, obtaining heads 5067 times in 10,000 attempts, for an average of 0.5067.

These results suggest that a coin can be a reasonably random instrument when the two possible results are equally balanced. If you wish to repeat these experiments, you will have to make sure that you catch the coin when it is still in the air – if you let the coin roll on the ground before it falls to one side, then the different designs on the two faces normally favor one side or the other.

If you count the relative numbers of zeros and ones in an imagined sequence you will be surprised. You will see that their sequence has an average that may be quite far from 0.5, while the sequence generated by coins is very close to this value. But where the human experimenters usually leave their mark is in the sequences of consecutive ones or zeros, the "runs". People think it is more realistic to create runs of ones or zeros that are short, resulting in frequent changes between the two numbers. Our intuition tells us that if we toss coins we are more likely to obtain the sequence 0100101101, as an example, than the sequence 0000011111, to give another example. But in reality both these sequences are equally likely to occur, though patterns similar to the first example are more frequent than those similar to the second. Human beings have difficulty in intuitively estimating the length and number of runs of the same digit.

If you have taken the time to create a sequence of 250 digits, now count the number of runs with three or more identical digits. It is very likely that you will have created very few, as your intuition has likely told you that such runs are not very likely. However, in the real world, a completely random sequence of ones and zeros (that is, the zeros and ones occur with equal probability and independently of the values that have already occurred) should have about thirty-two runs of three or more identical digits. Was your sequence like that? Quite frankly, it is highly unlikely. And what happens with runs of four or more identical digits? Does your sequence contain any of them? Perhaps it will surprise you to know there should be about sixteen. And what about sequences of five or more identical elements? Again, you probably didn't predict that there should be about eight, just as there ought to be four sequences of six digits, and two with at least seven elements.

If you now take the trouble to toss a coin 250 times, you will find that the runs described above will occur at, or very close to, the frequencies stated here. If you prefer to toss the coin fewer times to create a shorter sequence, you can compare your self-created sequence with that produced by your coin tossing experiment. Now that you know people tend to avoid runs of one number, as they associate alternating numbers with randomness and runs with a deliberate pattern, perhaps you are no longer the ideal person to do this experiment. Persuade a friend to join you in the experiment, asking them to write a random list of zeros and ones. Then toss the coin once for each of the zeros and ones on the list and count the number of runs. You will undoubtedly see that the coin tosses have produced more runs than your friend's list contains.

All of this may seem like a trivial game, but the reality is that creating a series of random numbers is a very important matter in science. For instance, in statistics it is known that random samples have very desirable properties that make them excellent candidates to "represent" a population. When organizing experiments, for example, to test drugs that are administered to one group of people but not to another group (who form the control group), it is important to know how to make a random selection of the elements that form part of each group, in order to avoid bias due to the subjectivity of the analyst. In computational science and in all areas that use computer simulations, it is important to make use of random numbers that allow algorithms to simulate the variability of real processes. In all these cases, the objective is to obtain sequences that have properties similar to those of sequences resulting from the toss of a coin, and that avoid the inherent subjectivity of human evaluation.

In the past, scientists made use of tables of random numbers that were expressly created for this purpose via very laborious procedures. Nowadays, though, these random numbers are obtained from a computer, by means of algebraic processes that produce random numerical sequences. In reality, these numbers are generated by deterministic processes, so they are "pseudo-random" numbers. But, as in chaos, the sequences obtained are such good reproductions of completely random sequences that they pass all possible tests, and so can be considered random for all intents and purposes. As the mathematician Donald Knuth, one of today's most renowned computational scientists, said, "random numbers should not be generated with a method chosen at random. Some theory should be used".[1] In other words, chance is too important to be left to chance.

---

[1] D. E. Knuth, *The Art of Computer Programming,* vol. II, Addison-Wesley, 1969, p. 6. The italics are in the original.

CONJECTURES AND PROOFS

The so-called Collatz conjecture was formulated in 1937 by the German mathematician Lothar Collatz. A conjecture concerns a mathematical assumption, something that we think is true, but that has never been finally proved or disproved. Like some of the most famous mathematical assumptions, these conjectures are easy to understand and they pass the common sense test, but they are tremendously difficult to prove or disprove.

Collatz' conjecture states that the number 1 is always obtained after making certain sequential operations starting from any natural number (positive integer). It works this way. We start with a positive integer. If this number is even, divide it by 2. If it is odd, multiply it by 3 and add 1. At the end of this process, a new number is produced. We take that number and repeat the whole process. It was Collatz's conjecture that if this sequence of operations were repeated indefinitely, the final result would inevitably be 1.

There is nothing to match an example. Let's start with 6. As it is even, divide it by 2 to obtain 3. As this number is odd, multiply it by 3, then add 1, and the result is 10. Continue this process... If you got your math right, you will find that the numbers obtained are: 6, 3, 10, 5, 16, 8, 4, 2, 1. We do end up with 1, after all. You can try this with other numbers, but you will almost certainly end up with 1, as countless others have tried this and always ended up with 1.

The Portuguese scientist Tomás Oliveira e Silva explored a great number of hypotheses, starting with the number 1 and getting past the number 27 thousand trillion. He did not come up with one case in which

N. Crato, *Figuring It Out*, DOI 10.1007/978-3-642-04833-3_40,
© Springer-Verlag Berlin Heidelberg 2010

he did not end up with the number 1. While this is an important result, it is not conclusive enough for mathematicians. It may well still be possible to find a number that has not yet been checked, and that would disprove the conjecture. In the absence of a rigorous proof or an example that refutes the conjecture, we cannot be sure whether the conjecture is definitively true or false.

Mr. Benford

Would you like to win a bet? Let's see if you can manage to guess the first digit of a number or at least whether that number is less than 4. Tell somebody to think of everyday-life numbers, such as the street number of a house or the amount of money they have in their checking account, or a percentage rate they commonly use such as a mortgage interest rate, or, perhaps, a physical or mathematical constant such as pi. If your bet is that this number begins with 1, 2, or 3, you will be right more than half the time, in fact, you will be right 60.2% of the time. Your betting partner will no doubt be surprised by your success rate, for at first sight there is no reason at all for the laws of chance to produce such a result.

Many people, even those who know probability theory quite well, are often still amazed by this fact, and cannot see any mathematical basis for such a deviation from common sense. In reality, any number can start with any one of the digits from 1 to 9 (we exclude 0, as it not the first significant digit of any number). But will any one of these nine digits have the same probable frequency? Why would a number that is selected at random be more likely to start with 1, 2, or 3? Shouldn't there be a uniform probability distribution, as they say, with a probability of 1/9 for each digit? And isn't 3/9 less than half, which would mean our wager should be a losing bet?

In reality the first digits of the numbers we encounter in everyday life do not all occur with the same probability. Number 1 is the most frequent, about 30.1% of the time, followed by number 2, with a 17.6% probability of occurrence, and so on, with the probability decreasing systematically until number 9, which only occurs in 4.6% of cases.

N. Crato, *Figuring It Out*, DOI 10.1007/978-3-642-04833-3_41,

The first person to discover this strange fact appears to have been the American astronomer Simon Newcomb, who published his findings in 1881. But the first to make a thorough study of the phenomenon was the American physicist Frank Benford, who worked for General Electric. He published his findings in 1938, in the periodical *Proceedings of the American Philosophical Society* (78, 551–572), but his work never became as well-known as it deserved. William Feller (1906–1970), who was a professor at Princeton for many years, was one of the few who recognized the significance of Benford's work, which he noted in his famous manual on probability theory.[1] Benford had noticed that the first pages of the logarithmic tables were much more worn than subsequent pages, meaning that pages with tables for numbers starting with 1 were more worn than the other pages. The tables seemed to be used less and less as they progressed to numbers that started with higher digits.

Nowadays, we rarely use logarithmic tables, as calculators and computers work much more precisely and rapidly. But these tables were used intensively over a period of several 100 years, from the start of the 17th century, when logarithms were created and popularized by the Scottish mathematician John Napier (1550–1617), until just a few decades ago, when electronic calculators first appeared. While those tables may have fallen into disuse, logarithms are still used extensively in the fields of mathematics and computation.

Surprised by his discovery, Benford wondered whether this could also happen in other circumstances, so he analyzed a vast variety of numbers, including the surface areas of rivers, sports statistics, numbers published in the press, and the addresses of celebrities, all in all, more than 20,000 sets of numbers. It turned out that in each case he encountered the same "anomaly", as he called it, verifying that more than 30% of the numbers started with 1, whereas barely 5% began with 9. What was the explanation?

---

[1]   W. Feller, *An Introduction to Probability Theory and its Applications*, Vol. 2, 3rd edition, Wiley, New York, NY, 1968

The phenomenon is difficult to comprehend, but there are some examples that help make it clearer. For instance, let's take the house numbers in a street address. How many houses are there on average in a street? Exactly 999? That's not very likely. Exactly 99? Not very probable either. So let's agree on 50 as an example. In this case, it is quite easy to work out how many house numbers begin with 1. There are eleven of them: house number 1, then 10, 11, 12, and so on until 19. The probability that a house chosen at random will have a number starting with 1 is 11/50 or 22%. On the other hand, there would only be one house with a number beginning with 9, i.e., house number 9 itself. So the probability of a house having a number starting with 9 is 1/50, which is only 2%.

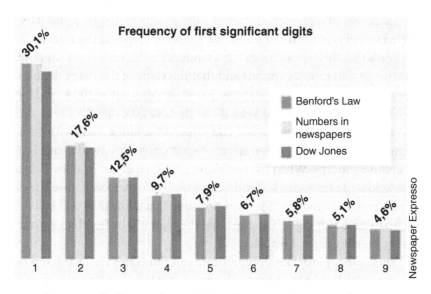

According to Benford's Law, the probability for a digit k to be the first significant digit in a number is not 1/9, but the log to the base 10 of the quantity $(k + 1)/k$. Therefore the probability that 1 is the first significant digit is about 30.1%, whereas the probability for 9 is only about 4.6%. The logarithm is an exponent for a given base number, and any positive number can be expressed as a power of a specific base number. For example, 1000 can also be expressed as 10^3, so its logarithm to the base 10 is 3.

The bar chart compares the frequencies of the first significant digits forecast by Benford's Law with those observed on the first pages of newspapers and with the values in the Dow Jones stock index. There is great correlation between the forecast frequencies and those actually observed.

If we think about it carefully, for any one of the nine digits 0–9 to have the same probability of being the first significant digit in a door number, the street has to have exactly 9 houses or 99, or 999 . . . which certainly will not occur very often. In any other context there is a bias against the higher digits, while the lower digits appear more frequently.

Another clear example is provided by growth rates. Just think of a stock exchange index, or a price index, or a production index. When the index was first created, it started with a value of 100 (it could have started with any other number, but the problem would still be identical). We know that this type of index has a tendency to increase. Let's suppose that the growth rate is constant and that the value of the index doubles every year. During its first year, the index remains in the 100 s, and by the beginning of the second year, it has risen to 200, then to 400 at the start of the third year, and so on. The value of the index starts with 1 for 12 months. Then the periods during which it starts with 2 and 3 each last for about 6 months. When the first significant digit at last reaches 9, this only holds true for about 1 month! It is easy to see that, even if we select a growth rate at random, it is most probable that a low digit will still be the first significant digit. If you bet that the digit will be less than 4, your chance of being right is 60.2%.

This case is particularly significant, as it illustrates the fact that it is an exponential and not a linear behavior that is the dominating factor in many natural and social phenomena. The behavior of the first significant digits reflects this fact.

Frank Benford understood this, and formulated what he called the "law of anomalous numbers", today usually referred to as Benford's Law. Although his formulation was ignored for many years, it is once more attracting the attention of mathematicians and statisticians. As often happens in mathematics, it is sometimes only years later that the possible applications and relevance of some results are seen. Benford's Law, for example, is beginning to be applied in a surprising domain, the detection of tax fraud. Mark Nigrini, an accountant with a statistical background who worked in Cape Town, New York, Kansas, and Texas before becoming a professor in The College of New Jersey, has created a method for analyzing tax declarations by checking the frequency of occurrence of the first significant digits.

People who systematically falsify a tax declaration by inventing non-existent expenses try to use numbers that seem to be as completely random and different as possible. Tax officials may have their suspicions if, for example, exactly the same quantity of expenses is claimed for each business trip. Well, the smartest cheats try to make the numbers as uniformly random as possible, starting as often with a 9 as with a 1. It is at this point that the new technique pioneered by Mark Nigrini comes into play. Many of the tests he has devised are quite simple; they only count the frequency of occurrence of the first significant digits. Some of the other tests are more complex.

By the way, if you are thinking about using Benford's Law to help you find winning numbers in the lottery, forget it. The distribution of lottery numbers is uniform, and does not follow Benford's Law. But you can certainly use it to win bets with your friends.

FINANCIAL FRACTALS

The mathematician Benoît Mandelbrot once said that his childhood ambition was to become the "Newton" of a specific mathematical field, no matter how small, in other words to propose something creative and completely innovative in a particular area of mathematics. Ultimately he chose what came to be called fractal objects, at that time relatively unknown geometrical figures with curious properties.

It is very difficult, if not impossible, to define fractals in non-technical terms, but the basic idea is that such objects contain a structure that is reproduced at every scale. For example, a square is not a fractal object – if part of one of a square's edges is amplified, what we see is a segment of a completely smooth straight line without any of the outlines that characterize a square. But a snowflake can have a fractal structure. When its branch-like aspects are observed under a microscope, we see within them other, similar branch-like structures. If we observe these through an even more powerful microscope, we can discern even more similar structures, and this pattern can go on and on. The snowflake has a structure that repeats itself at various scales. Clearly, this degree of similarity is not present at the molecular scale, but we can imagine geometric models in which patterns are repeated infinitely.

Mandelbrot called such objects fractals and he recognized that their dimension, at least in a certain technical sense, could have a fractional value. A segment of a straight line has one dimension, a square has two dimensions, a cube three dimensions. But objects such as snowflakes can have a fractional dimension, such as two and a half.

N. Crato, *Figuring It Out*, DOI 10.1007/978-3-642-04833-3_42,
© Springer-Verlag Berlin Heidelberg 2010

Mandelbrot offers fascinating examples of the strange properties of fractal objects measured at different scales. For example, what is the length of the coastline of Great Britain? If we use a good map to measure it, we will find a numerical result; but if we then use an even more detailed map, we will count more bays and capes, so the result will be a larger number than before; if we then decide to measure the rocks and coves, the number increases even further.

Mandelbrot concluded that the measurement of the length of certain irregular objects depends on the gauge that is used. The smaller and more precise the gauge, the greater the length that is measured. The border between Spain and Portugal is shown differently on Spanish and on Portuguese maps. When measured and officially documented by the Portuguese it is almost 20% longer than according to the Spanish records. This is also the case with the border between Belgium and the Netherlands, for example, where the difference is also about 20%, with the smaller country always being the one that reports the larger value. Mandelbrot has an explanation for this: the smaller country measures its border more carefully, using a more precise gauge. The length of the border depends on the scale that is used.

For more than 40 years, Mandelbrot has consistently emphasized that the phenomena of scale and of self-reproduction are equally applicable in terms of economic and financial matters as well. The setting of prices is the result of various individual choices, decisions made by many buyers and sellers. And the variation over a long period is the result of many shorter-term variations. Large movements result from small movements, which in turn result from even smaller movements, all of which have similar properties. They are the snowflakes whirled around by the economy.

Mandelbrot derived two properties as the mathematical result of this hypothesis, which he metaphorically called the "Noah effect" and the "Joseph effect". By referencing the Biblical flood, Mandelbrot asserts that extremely large variations may be produced occasionally, i.e., very extreme values occur with a significant probability. Using the seven fat years and the seven lean years Joseph experienced in Egypt as a reference point, Mandelbrot asserts that economic phenomena exhibit a degree of persistence that is technically known as "long-term memory", prolonging for long periods the effects of "shocks" or irregular variations.

What is absolutely amazing about this is that these two basic ideas, which were not taken seriously at the time, have come to dominate much of the econometric debate today.

The extreme irregularity of financial series, the drastic collapses such as occurred on "Black Monday" in New York in October 1987, and also more recently at the start of the current financial crisis, and the subsequent spectacular recoveries, are all phenomena that are difficult to explain by means of the probability distributions normally used in econometrics, like the normal or Gaussian curve. Extreme movements seem to indicate that the financial series do not have a finite variance or volatility, as it can be technically called. Even today this problem is a point for debate between economists and statisticians: every year dozens and dozens of new studies are published, re-evaluating the hypotheses of infinite variance introduced by Mandelbrot.

On the other hand, the persistence of variability in financial series has been given more credence by recent econometric studies, something Mandelbrot foresaw more than 40 years ago, which is called "volatility persistence" or "contagious speculation". What happens here is apparently paradoxical. Essentially, it holds that movements in financial markets are impossible to foresee – nobody knows whether the stock market is going to rise or fall tomorrow. But we can make reasonable statistical forecasts for the magnitude of the changes, the so-called *volatility*. If the market is nervous, if it has experienced great fluctuations, then it is natural to predict that tomorrow will bring a major fluctuation. If the market is calm and stable, then it is likely that tomorrow it will remain calm, and that any price changes will be minor. High volatility tends to be persistent, as does low volatility. This "long-term memory" of volatility in financial markets, which represents the Joseph effect proposed by Mandelbrot more than 40 years ago, only began to be statistically evaluated in the 1990s, and it became an interesting model for the economic sciences. Clearly, Mandelbrot had much more important things to tell economists than many believed at the time.

TURING'S TEST

Will we succeed one day in constructing machines capable of thinking? This is a more profound question than it at first seems, and has been one of the most hotly debated topics among philosophers over the centuries. When computers first appeared, these debates flared up again and their tone became more pressing. Nowadays, with computer programs capable of beating the world's chess champion, it makes a lot of sense to ask whether, when all is said and done, artificial intelligence is not in fact true intelligence.

The modern debate about the possibility of creating intelligent machines started in 1950, when the first computers became available. In England the émigré Hungarian chemist Michael Polanyi (1891–1976) went to great lengths to show that the human mind could not be reduced to a simple mechanical system. He based his arguments on a theory propounded by Kurt Gödel (1906–1978), a mathematician and logician who had demonstrated that it is impossible to construct the whole field of mathematics on the basis of a fixed system of axioms. The philosopher Karl Popper (1902–1994) also contributed to the debate, affirming that only the human brain can make sense of a machine's meaningless capacity to produce truth. In other words, computers cannot see any significance in any true propositions that they may construct, as only the human mind can make sense of the instructions provided to the machines and to the results of their operations.

At this juncture the debate was joined by Alan Turing (1912–1954), the British mathematician who had managed to break the German encoding machine during the Second World War. In 1937 Turing had shown that there are mathematical problems that are impossible to resolve by means of automatic computation, and accordingly he did not believe in the absolute power of the computer. Nor, however, did he believe in the non-material nature of the mind. In Turing's view, intelligence only exists and becomes manifest in dialog and interaction with the environment, most commonly among people. That is, comparison with the self is the only basis a human being can use to evaluate whether a person or a machine is, or is not, intelligent.

Turing suggested a practical test to characterize intelligence. In a study published in 1950 in the philosophical journal *Mind*, he imagined a game in which a person communicated in writing with another person or with a machine. Using a series of questions, the inquirer had to try to find out whether he or she was writing to a human or a machine. A machine's intelligence would be measured by the extent to which it managed to fool the inquirer. So, for Turing, a computer could be intelligent to the extent that humans do not unmask it.

Turing's test had an enormous influence on the development of a branch of computational science known as artificial intelligence. First specialists tried to understand the way the human brain functioned by reproducing it, as far as was possible, in a computer. Later they perceived that there are some areas in which machines are much superior to humans, and others in which humans defeat machines emphatically. They also observed that these two aspects can be differentiated as follows: whereas human beings are capable of distinguishing or creating patterns based on a variety of information, machines are capable of systematic and repetitive searches. So improvements made to the capacity of computers would result from algorithmic improvements for the systematic search for solutions.

When the Internet arrived, the distinction between human intelligence and artificial intelligence became a pressing problem that had to be resolved in practice. Some computer programs had infiltrated into *chats* on Yahoo and other systems, collecting information on the human participants and inserting commercial references to companies or products. Other programs had created problems for the major portals by automatically registering thousands of non-existent users, thereby creating innumerable email accounts that were then used to send *spam*, massive commercial mail shots that slow down servers and accumulate in our mailboxes, while others have succeeded in worming their way into individual accounts, and trying out keywords repeatedly until they gain access to the account. The security systems devised to prevent people from violating systems can be rendered powerless when confronted by machines.

To resolve this problem, technicians at Yahoo set up a cooperative program with scientists at the Carnegie Mellon University in Pittsburgh. The so-called CAPTCHA project (www.captcha.net) attempts to generate tests that only human beings can easily pass, and that computers usually fail. Some of these tests are based on the distortion of words or phrases, which OCR (optical character recognition) programs find difficult to decipher. The ability to recognize visual patterns is the result of an evolution spanning millions of years and of very long training in the case of each individual. It is not easy to figure out exactly how this capability developed, much less reduce it to a set of instructions that can be imparted to a computer. However, it is simple to write a computer program for operations that we were taught rationally, such as logical or arithmetical functions, which is why machines can easily handle problems that can be reduced to these formal operations, though they cannot recognize subtler patterns in messages containing a lot of visual noise.

Various companies are now engaged in exploring this limitation of computers in order to find ways of protecting their own services. Yahoo's CAPTCHA project involves a process that restricts the automatic registration of users. If you want to open an email account in this portal (www.yahoo.com), then at one stage of the procedure you will have to read some letters written in a bizarre script and then type the letters into the registration form. If you don't read the letters correctly (and OCR programs have great difficulty with this), you won't be given access to the next stage of the procedure, and so you will not be permitted to open the email account. In this case human beings succeed in doing something machines can't do.

The CAPTCHA concept is now widely used. Turing's test gave rise to a flourishing investigation involving various applications in the computer industry. Because of this, today we can provide multiple responses to the question: "Do machines think?" Fortunately for us, there are still occasions when they don't know how to think.

DNA COMPUTERS

On April 25, 1953, James D. Watson and Francis Crick published an article in the journal *Nature* that was less than two pages long. In this study they presented the famous double-helix structure of DNA, and as a consequence the world today has changed. The discovery of DNA has opened up new paths in biology, medicine, agriculture, forensic science, and in numerous other technical and scientific fields. The double helix has appeared in works of art, and the abbreviation has even been adopted as the name for perfumes. And now something that people have been talking about for some time seems to be coming true: DNA could begin to be used for computation. In the near future PCs might no longer use silicone, but instead an aqueous solution of DNA molecules.

In 1994 Leonard M. Adleman was the first person to show how DNA could be used to resolve computational problems. The problem he solved using DNA could today be solved by a computer in a fraction of a second. It took Adleman 7 days to accomplish, but in the process he demonstrated how DNA could be used as a basis for calculation.

Adleman was faced with a "traveling salesman" problem. He imagined seven cities and a salesman who had to visit each one of them, starting from a specific city, and never visiting any city twice. This might seem simple, but when we increase the number of cities the problem rapidly becomes impossible to tackle. We may be forced to investigate each possible solution, which would require tremendous computational resources. This method came to be called the "brute force method", and is the one Adleman employed.

N. Crato, *Figuring It Out*, DOI 10.1007/978-3-642-04833-3_44,
© Springer-Verlag Berlin Heidelberg 2010

First Adleman generated all the possible routes. He then isolated those that originated in the designated city. He subsequently coordinated the itineraries with the exact number of cities. And finally he selected the paths that only visited each city once. To do all that, he made use of the way in which DNA transmits information.

Whereas computers store data in sequences of zeros and ones, DNA codes them in sequences of four bases, represented by the letters A, T, C and G, aligned sequentially in chains. The double helix is formed by two complementary chains that are joined in such a way that A on one chain is linked to T on the other, just as C is linked to G; the two chains are connected via bonds between these pairs. For example, this means that the complement of the sequence ATCAG is TAGTC. One chain can only be associated ("hybridized") with another if they both have complementary sequences.

Adleman encoded each city as a particular sequence of these four letters. He then did the same for the connections between the cities. For example, Denver could be represented by CTACGG and Salt Lake City by ATGCCG. The route connecting these two cities could be CGGATG, combining the last three letters of Denver with the first three letters of Salt Lake City. The complement of this route sequence is GCCTAC, which could be associated with the sequences for the two cities, as the letter G is associated with C (the fourth letter of Denver), C with G (the fifth letter of that city), and so on until the last letter, C, which is associated with G (the third letter of Salt Lake City). In other words, within a long DNA sequence, the appearance of GCCTAC is associated with a sequence containing Denver followed by Salt Lake City.

Adleman then constructed DNA fragments containing the codes of the cities in random order, which is a relatively simple task using a DNA synthesizer. He also constructed sequences that contained the codes for the connections between the cities, also in random order. Then, using the PCR (polymer chain reaction) technique that has since been completely mastered, he produced many copies of the sequence of the connections that began with the departure city, such as Denver, and ended with the destination city, for example Santa Fe.

Next, Adleman used another well-known technique to select DNA fragments that contained exactly 30 bases (5 cities × 6 bases) in such a way that none of the selected routes passed through any city twice. Finally he used a well-known chemical technique (affinity purification) to review the residual groupings and select the molecules that contained all the cities. The final groupings, linking the chains of cities to those of the connections, constitute a solution, or solutions, to the problem.

The problem that Adleman solved was just an initial demonstration of the computational power of DNA, but since that time progress has been rapid. In 2003, a team from the Weizmann Institute in Israel succeeded in creating a system in which propulsion energy itself is provided by the DNA molecule. The *Guinness Book of World Records* recognized this system as the world's smallest biological computational device to date.

Doubts have been expressed about the feasibility of creating operational computers based on DNA: the calculations are performed rapidly, but the preparation and reading processes are very slow. Nevertheless, this miraculous molecule has alluring properties. It possesses a capacity for storing information that electronic systems can only dream of: a single cubic centimeter (about one-sixteenth of a cubic inch) of DNA can store as much data as a trillion CDs. Even more importantly, it is a multi-processor: the molecules work on parallel tracks, and so are able to produce millions of answers simultaneously, and thus incredibly quickly. For the problems that it can resolve, this system first constructed in Israel is already one hundred thousand times faster than today's PCs.

MAGICAL MULTIPLICATION

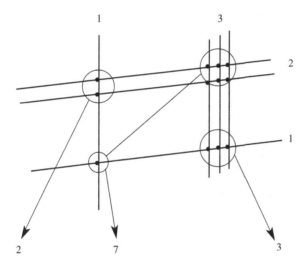

The video shows what it claims is a new method of multiplying. It is simple, somewhat strange, but it always seems to work. It starts by showing us how to multiply 21 by 13. To do this we have to draw two horizontal lines, which represent the number of tens (2) in the first number. Below we draw a single line that represents the number of units (1) it contains. Then we draw the second number using vertical lines: one on the left for the number of tens (1), and three on the right for the number of units (3).

Then we count the points at which the vertical and horizontal lines intersect. There are four corners, each with its points. At the top left

N. Crato, *Figuring It Out*, DOI 10.1007/978-3-642-04833-3_45,
© Springer-Verlag Berlin Heidelberg 2010

there are two intersecting points, so we write a 2 for the hundreds. At the bottom right there are three intersection points, so we write 3 for the units. The remaining two corners have a total of seven points, so we write 7 for the tens. That's it: $21 \times 13 = 273$.

The video continues with a more complicated example to show that the method always works: $123 \times 321 = 39{,}483$. In this case there are five types of intersection points that provide the five digits in the answer. There is one more step that has to be taken: "and carry over", as in the usual multiplication algorithm.

In fact, strange as it may seem, these drawn lines simply represent a method that is similar to the way we normally multiply two numbers. In the end, what is 21×13? We just have to note that the numbers 21 and 13 simply represent our decimal notation for the quantities. So we see that $21 = 2 \times 10 + 1$ and that $13 = 1 \times 10 + 3$. The multiplication can be done with the two numbers so partitioned and adding the partial results ($20 \times 10 + 20 \times 3 + 1 \times 10 + 1 \times 3$). This is easy and it works. It must work! The intersection of x lines with y lines has to be $x \times y$ points. That is exactly what multiplication is.

Having arrived at this point, you may be tempted to ask: why didn't they teach me that in school? Wouldn't it have been easier and more fun? You are right: it would be easier and more fun, but only for very basic numbers. Try multiplying 99 by 99 and you will see that it is neither easy nor fun to arrive at the answer by drawing lines. In fact, you will find that it is easier and less error-prone to multiply these numbers as we were originally taught. The multiplication algorithm that we learned (or should have learned) in school is the result of a series of trials and improvements that have lasted for hundreds of years. It is worth taking the trouble to master it.

π Day

We all know certain commemorative dates by heart. In the U.S., for example, Mother's Day falls on the second Sunday in May, Father's Day on the third Sunday in June, Labor Day on the first Monday in September. And π day? Do you know when that is?

This date was not chosen at random. To commemorate the number 3.14, the 3rd month and the 14th day were chosen, so π day falls on March 14. In schools the date has been observed for some decades. But now the U.S. House of Representatives has made it official: on March 12, 2009 with 391 Yeas and only 10 Nays the House supported "the designation of a 'π Day' and its celebration around the world". The resolution further explained that "π can be approximated as 3.14, and thus March 14" is an appropriate day for such celebration.

Of course π is not exactly 3.14. It is a number with many decimal places. Some people write it as 3.1416. The more precise calculators note it as 3.1415926535. And the number does not stop at that point. As a matter of fact, where does it end?

It has been known for a long time that π is a number that does not have a finite decimal form. It cannot be written as a finite decimal number, nor as a periodical infinite decimal. It is an irrational number, which means it cannot be written as the ratio of two integers, so its decimal expression is endless and does not repeat itself. No matter how we write π, we always have to deal with an approximation.

An initial approximation to π was the number 3, as is implicit in the *Bible*, in the *Book of Kings*, written in about the 6th century BC. Even before that, the Babylonians had a better approximation, 3.125,

N. Crato, *Figuring It Out*, DOI 10.1007/978-3-642-04833-3_46,
© Springer-Verlag Berlin Heidelberg 2010

and the Egyptians implicitly used the square of 16/9, which is 3.16049... In the 3rd century BC Archimedes discovered a rigorous method for calculating π, and obtained an approximation that was correct to the second decimal place.

Archimedes's method entails fitting a circumference into inscribed and circumscribed polygons. The perimeters of the polygons are simple to calculate, and the perimeter of the circumference is fitted between them. It is an ingenious method and it allows us to obtain any desired approximation simply by using polygons with more sides.

During the Middle Ages Archimedes's work was rediscovered by the Arabs, and later by the Europeans. A race to calculate the value of π developed. In the 9th century al-Khwarizmi calculated it to 4 decimal places. By the end of the 16th century mathematicians had achieved twenty decimal places, and then by the close of the 18th century they had arrived at 140. By the beginning of the 20th century more than 500 decimal places had been achieved. Somewhat later computers made their appearance. Thousands of decimal places were calculated, then millions, and later thousands of millions... The record is currently held by the Japanese mathematician Yasumasa Kanada, who managed to obtain 1,241,100,000,000 digits by the end of 2002, which is more than a trillion decimal places.

Apart from having started a race to make an ever more precise calculation of the value of π, the method devised by Archimedes allowed this calculation to become autonomous from geometry. In fact, when we write the equations for the perimeters of the polygons, π appears as a pure number.

In the 17th century various mathematicians, including Leibniz and Newton, devised equations that were more efficient at calculating π, and it came to be observed that this number pops up in the most unexpected circumstances. The strangest process for calculating π, however, may well be the probabilistic method proposed by the French naturalist Count de Buffon (1707–1788), a curious fellow we have already encountered. You can try it out at home.

Begin by getting a sewing needle. Then draw a series of parallel straight lines right across a large sheet of paper. The straight lines must always be the same distance apart: to make things simple, separate them by twice the length of the needle. Then toss the needle repeatedly on to the paper. Count the number of times the needle intersects a straight line. Divide the total number of times you tossed the needle by the number of times it touched a straight line, and the result is an approximation of π. This method is probabilistic and converges very gradually. You may well need to toss the needle hundreds of times before you get the first decimal place correct. But the experiment is so strange and curious that it is worth trying it out. It is a good way of commemorating π day.

THE BEST JOB IN THE WORLD

Many scientists will secretly admit that they enjoy their jobs so much they would pay to do what they are paid for. Of course, this is something of an exaggeration. Very few of them could actually keep up a full-time job in a scientific field if they were not paid. The "good old days" when aristocrats pursued science as a hobby are long past. Today, science is a job. A good job.

In the United States in particular, careers in science are well remunerated. According to a survey published recently by CareerCast.com, a business organization, science is the top profession. That's right: the best.

This survey contradicts the disdain for studying and obtaining qualifications that is often to be read in public comments made thoughtlessly by people in responsible positions. Young people have to be reminded that studying is a precious investment in the future. That has to be stated repeatedly. A qualification in the sciences or the arts brings various professional benefits. It is not just a question of salary. It involves job quality, quality of life and cultural richness.

The CareerCast study presents a ranking list (with all the simplifications typical of a list of this type) along with a good deal of comparative information also extracted from the survey. In it, 200 types of jobs were reviewed and then classified according to five basic criteria: work environment, remuneration, future prospects, physical requirements and emotional stress.

The survey endeavors to use objective criteria. Job elements such as the necessity to adopt uncomfortable positions, put up with possibly

N. Crato, *Figuring It Out*, DOI 10.1007/978-3-642-04833-3_47,
© Springer-Verlag Berlin Heidelberg 2010

toxic conditions, or accept confinement in small spaces are among the data used to assess the work environment. The factors used to classify compensation include the dates paychecks are issued, and any additional bonuses or perks. Regarding future prospects, some of the criteria are the prospective risk of unemployment, the possibility of future salary increases, and chances of promotion.

The fourth criterion, physical requirements, is a domain loved by physicists, who have assisted official statisticians in measuring this area by taking various jobs and evaluating them on the basis of such factors as the weight that has to be lifted, types of movements required (kneeling, bending, etc.), and the work actually performed (in the physical sense of the word).

Finally, the study assesses emotional stress, considering such factors as relative competitiveness, risks, and pressure to respond.

After all these factors are considered, it turns out the worst job (200th in the list) according to this study, is lumberjacking, which involves tremendous physical risks for a low average pay of less than $32,000 a year. Other jobs at the bottom of the list include taxi-drivers (198), fishermen (197), garbage collectors (195) and fire fighters (182). At the other extreme, the top-rated jobs include meteorologists (15), physicists (13), accountants (10), systems analysts (6) and IT engineers (5). These are surpassed by biologists (4).

The best three jobs in the survey are statisticians (that is, math specialists in data processing and inference), actuaries (that is, specialists in the mathematical calculation of the risks involved in the insurance and investment businesses), and finally, in first place, mathematicians!

OUT OF THIS WORLD

ELECTORAL PARADOXES

Voting in elections is one of humanity's great achievements. No better method has yet been invented to achieve a system of government that guarantees liberty and progress. But would it be possible to invent a better one?

Democracies adopt a system usually known as "one man, one vote". This is overall a fair system, though it is not free of paradoxes or contradictions, as has been noted since ancient times. Apparently Pliny the Younger, a lawyer and magistrate of Ancient Rome (this name is used to distinguish him from his uncle Pliny the Elder) was the first to reveal some of the paradoxes involving voting. In the 2nd century BC only an elite group was permitted to vote, but their problems were largely the same as ours today. Electoral paradoxes first began to be discussed systematically in the 17th century, with the objective of devising a perfect and rational electoral system, but soon issues began to arise.

The French mathematician Jean-Charles de Borda (1733–1799) was the first person to study these electoral issues systematically. Regarding the electoral system as a means of aggregating opinions to determine the collective choice, he noted that different voting methods lead to different results. This became known as Borda's paradox and was a popular topic of discussion at the time, although no satisfactory resolution to the paradox emerged.

On June 16, 1770, Borda formally presented the problem to the Royal French Academy citing an example in which 21 voters had to choose from among 3 candidates. He took under consideration the relative preferences of each voter, that is the way in which each voter ranked

N. Crato, *Figuring It Out*, DOI 10.1007/978-3-642-04833-3_48,
© Springer-Verlag Berlin Heidelberg 2010

the candidates. What he noticed was that, surprisingly, it was possible to elect a candidate the majority of voters had ranked last. This only required that the votes for the other candidates were sufficiently split. Nowadays this phenomenon is referred to as the election of a "Condorcet loser", that is of a candidate who loses in direct comparison with all the others. In Borda's example candidate A would lose the election if his only opponent was candidate B, and would also be defeated if his only opponent was candidate C, but wins the election if both B and C are competing against him.

To resolve this paradox, Borda proposed a system that came to be known as the Borda count. Instead of the "one man, one vote" system, Borda gave each voter the opportunity to award points to each candidate. With a total of three points at their disposal, voters could assign two points to their preferred candidate, one point to their second-preference candidate, and zero points to the third candidate. The points would then be added up, and the candidate with the most points would win.

BORDA'S PARADOX

Preference	1 Voter	7 Voters	7 Voters	6 Voters
1st choice	A	A	B	C
2nd choice	B	C	C	B
3rd choice	C	B	A	A

Each preference profile is shown in one of the columns together with the number of voters who chose it. So, for example, only one person voted for candidates A, B and C as their first, second and third choices respectively. In the second column we see that 7 voters ranked the candidates in the order A, C, B. In this example provided by Borda, the candidate with the most votes according to the one-man-one-vote system is A with 8 votes, compared with 7 for B and 6 for C. Nevertheless, A is the candidate most disliked by the majority of the electorate, since 13 out of 21 voters place him last in their ranking.

This system at first appears perfect, but it also has certain problems. First, why do the points awarded to each preference have to be successive whole numbers? Giving zero points to the least liked candidate, one point to the next candidate, and so on, might not result in an exact reflection of the voters' preferences. Why couldn't a voter give zero points to one candidate, half a point to another, and one and a half points to the preferred candidate? The curious thing about this is that under such a system it would be possible for a different candidate to win. Even if the ranking of preferences remained identical, which candidate wins still hinges on the weighting system being used. In other words, the Borda count system will not always produce the same result.

As another French mathematician and philosopher, the Marquis de Condorcet (1743–1794), showed, an even more serious electoral problem is that it is not always possible to aggregate the voters' preferences in a coherent manner. The ranking of the preferences of each voter should be *transitive:* if a voter ranks candidate A ahead of B, and B ahead of C, then he or she will also rank A ahead of C. However, this is not necessarily reflected in the total number of votes in an election. More people might vote for A than B, or for B than C, but still prefer C to A! How can this problem be resolved?

CONDORCET'S PARADOX

Preference	Group 1	Group 2	Group 3
1st choice	A	B	C
2nd choice	B	C	A
3rd choice	C	A	B

The first group of voters prefers candidate A, followed by B and then by C. There is a transitive logic in the choices expressed by this group. If A is preferred to B and B is preferred to C, then A is also preferred to C. But this transitivity property is not passed to the final election results. Let us suppose that each group is comprised of an identical number of voters. Then A will beat B, as this is the ranking preferred by groups 1 and 3. This is the majority opinion of the voters. On the other hand, B will also beat C, as this is the ranking preferred by groups 1 and 2. It would seem logical that, as A beats B and B beats C, then A should also beat C. Nevertheless C beats A, as you will see using the same reasoning: both group 2 and group 3 rank C ahead of A.

Donald Saari, a mathematician from the University of California at Irvine who has devoted himself to the study of electoral problems, has illustrated how minor changes in any electoral system can lead to major changes in election results. Saari is among the mathematicians and political scientists who have studied problems related to "public choice", a field that experienced major developments in the second half of the 20th century.

At this stage you may already suspect that a perfect system is very hard to come by. But the problem is even more complicated than it appears. Kenneth J. Arrow, a Nobel-winning American mathematician and economist, examined a set of electoral conditions that were apparently reasonable, such as the transitivity of preferences mentioned above, and demonstrated that it is not possible to have a single electoral system that satisfies all these conditions at the same time.

So what can be done? Mathematically, there is no solution to this problem, but society does not really need systems that are perfect. Rather, it needs rules that lead to choices that are acceptable to society, even if they are fallible and approximate. Mathematics can help people to perceive the problems of various electoral systems, but it does not go so far as to question the democratic system itself, as this is a social moral choice validated by history.

THE MELON PARADOX

This is a curious problem that comes up regularly in math competitions. You will see it in one form or another in published collections of problems from these competitions. It's not that the math itself is difficult. The difficulty lies in believing the results. As an example, let's start with a melon weighing 50 ounces. Only 1% of the mass of the melon is made up of solid matter, while the remaining 99% is water. The melon is left in the sun and dehydrates to such an extent that it now only contains 98% water. The question is: how much does the melon weigh now? The answer is easy, provided you do your sums properly. But let's start by guessing the weight. Will it be about 49 ounces? Or even 49 and a half ounces? Or just 45 ounces?

Most people do guess somewhere in the region of these numbers, perhaps figuring that 50 ounces × 98/99 is the solution, which would mean the melon weighs 49 and 49/99 ounces, which we can round off to 49 and 1/2 ounces. The surprise though, is none of these answers are correct. They are, in fact, completely wrong! It turns out the melon has lost half its weight, and now weighs just 25 ounces!

The calculation is simple: if 99% of the original melon was water, the solid content was 1/100th of 50 ounces. If the water content of the dehydrated melon was then reduced to 98%, the solid part is now 2%. This means that the total weight is now 50 times the weight of the solid content, which did not change. Therefore the weight has dropped to 25 ounces. It has been reduced by half. This problem is

N. Crato, *Figuring It Out*, DOI 10.1007/978-3-642-04833-3_49,
© Springer-Verlag Berlin Heidelberg 2010

educational, and almost a paradox. It shows how misleading intuition and simple proportions can be when we are dealing with relative measurements. But to be surprised, you have to start out by guessing the answer.

THE CUPCAKE PARADOX

When a fair-minded group of friends shares a plate of cupcakes, each one takes one and eats it, taking care to leave a cupcake for the next person. However, if the cupcakes are especially tasty and everyone is hungrier than usual, what happens when there is only one cupcake left on the plate? Carefully, one of the friends cuts the cake in half and takes one of the halves. A second person can't resist, and cuts the remaining half in half. Then a third friend comes forward and cuts the remainder in half. And so it goes on ... Theoretically we could imagine a virtual cake that is infinitely divisible, and a group of friends with all the time in the world to go on eating half of whatever was left of the cake.

At the end of time would all the cake be eaten? That is, does the sum $1/2 + 1/4 + 1/8 + \ldots$ equal 1? It is not difficult to see that the answer is yes, but it is easier to deduce this by subtraction than by addition. In other words, it is easier to see that what remains tends towards zero. In sums like this, calculators are not really much help.

Try out another problem that will make you aware of the limitations of the calculator. See if you can find out where this infinite sum converges: $1/2 + 1/3 + 1/4 + \ldots$ A reasonably accurate programmable calculator will make about one hundred million calculations and end up with a value in the vicinity of 18. No matter how good the numerical calculator is, the answer will always end up around a relatively small finite value. Nevertheless, this series actually does not have a fixed total; it increases infinitely. See if you can regroup these fractions and compare them with $1/2 + 1/2 + 1/2 + \ldots$ Obviously this sum is infinite. It will not stop at 18.

N. Crato, *Figuring It Out*, DOI 10.1007/978-3-642-04833-3_50,
© Springer-Verlag Berlin Heidelberg 2010

Provided you add a sufficient number of fractions, you will end up with a number greater than a million, or a billion, or a trillion trillions... whatever you like. However, after a certain time a calculator stops at a certain number, depending on the calculator, and this number will not increase no matter how many more fractions you add. To fully appreciate the problem, imagine that the calculator operates with a five-digit internal precision. In that case, if the calculator added 10–0.0001, it would not come up with 10.0001, but would continue to record the answer as 10 or 10.000, as it could not distinguish between 10.0001 and 10.000. In fact, calculators do have greater precision, but this problem always remains. It is this limitation that prevents the total from properly increasing when very small fractions are added; the machine just does not know any other way to do it.

Why don't you try another example? Enter 10 on your calculator. Press the square root key and then the square key. The answer will be 10, as you expected. But now enter 10, press the square root key 25 times and then press the square key 25 times. You will see that the answer is not 10, but something like 9.99239... depending on the precision of your calculator. The error is minute, but if you press the keys 33 times instead of 25 times, the answer you get should be 5.5732... which differs markedly from the original 10. It turns out that when you multiply rounded-off numbers over and over again, the result can be a disaster.

INFINITY

Galileo, whose scientific activities were celebrated during the International Year of Astronomy, considered various paradoxes having to do with infinity. One of the simplest and most illustrative paradoxes concerns two sets, one of the natural numbers $(1, 2, 3, \ldots)$, and one of their doubles $(2, 4, 6, \ldots)$. We can establish a one-to-one (bijection) correspondence between the two sets: 1 corresponds to 2, 2 corresponds to 4, 3 corresponds to 6, and so on. The first set seems to contain twice as many elements as the second set, because it contains both odd and even numbers. But doesn't the fact that we can establish a one-to-one correspondence between each number and its double indicate that each set has the same number of elements?

Seemingly, this was how human beings learned to count. Thousands of years ago, before writing had been invented, people counted sheep, or whatever, by collecting as many pebbles, or making as many notches, as there were sheep. The one-to-one correspondence between sheep and pebbles ensured that the two sets had the same number of elements.

It then seems clear that the sets $\{1, 2, 3, \ldots\}$ and $\{2, 4, 6, \ldots\}$ both have the same number of elements. At the same time, it seems clear that the first has twice the number of elements. . .

From this paradox, Galileo (1638) concluded that "the attributes of greater, lesser, and equal do not suit infinities, of which it cannot be said that one is greater, or less than, or equal to another".[1]

---

[1] G. Galilei, *Two New Sciences*, translated and edited by Stillman Drake. University of Wisconsin Press, Madison, WI, 1974, p. 40.

N. Crato, *Figuring It Out*, DOI 10.1007/978-3-642-04833-3_51,
© Springer-Verlag Berlin Heidelberg 2010

Two hundred and fifty years later, the German mathematician Julius Dedekind (1831–1916) took this idea as his starting point to define mathematical infinity. According to Dedekind, a set is infinite if it is possible to establish a one-to-one correspondence between it and one of its subsets. This is so in the case of the correspondence between the set of natural numbers and the subset of even numbers. Even numbers are also natural numbers, but there are natural numbers that are not even numbers, as is well known.

One of the most delightful paradoxes concerning infinity is "Hilbert's Hotel", usually attributed to the German mathematician David Hilbert, though it is more likely a product of physicist George Gamow's imagination. Gamow was the first person to chronicle "Hilbert's Hotel" in writing. It goes like this: imagine that a hotel with an infinite number of rooms is fully booked. In this imaginary hotel, it is always possible to make room for an additional guest. This is accomplished by the receptionist simply asking the guest in room 1 to move to room 2, the guest in room 2 to move to room 3, and so on. The guest who has just arrived is given room 1 and nobody is left out. But if there is always room for one more guest, then there is also always room for two more guests. And if there is always room for two more, there is always room for three more. The clients only have to move to a room with the corresponding number. Having an infinite number of guests arrive at the same time makes it more difficult, but if you think about it I am sure you will find a solution.

Hilbert's hotels would be ideal for a hotel chain. Not so for the guests. It cannot be pleasant to spend all night moving from one room to another.

Unfair Games

Imagine that we are in a casino that is promoting the following game: We put 100 dollars on the table, and win or lose by tossing a coin. If it is heads, we win 40 dollars, and if it is tails we lose 30 dollars. Should we join the game?

If the coin being used is perfectly balanced and tossed correctly, heads and tails have the same probability of coming up. This means we have a 1 in 2 chance of winning 40 dollars and a 1 in 2 chance of losing 30 dollars if we join the game. The expected value of this game is 40/2 – 30/2, or 5 dollars. This means that if we put 100 dollars on the table many times and play the game for a long time, we will win approximately 5 dollars each time we toss the coin. After tossing the coin one thousand times we should have won about 5000 dollars. This means it would pay us to go to such a casino. It would be like having our own printing press for dollar bills.

But the croupier knows how annoying it is to have to keep on putting 100 dollar bills on the table, and decides to simplify the game. Now instead of winning 40% or losing 30% of 100 dollars each time we toss the coin, we have to place 100 dollars on the table and then we can play the game, but this time we win 40% or lose 30% of whatever sum is on the table. For example, if the results of three tosses are heads, again heads, and then tails, our 100 dollars at first increase by 40% of 100–140, then by another 40%, now of 140, making the total 196. Following the third toss, that 196 is reduced by 30%, leaving 137 dollars and 20 cents on the table.

N. Crato, *Figuring It Out*, DOI 10.1007/978-3-642-04833-3_52,
© Springer-Verlag Berlin Heidelberg 2010

It does seem as if the croupier is making things simpler for us. We start the printing press rolling and enjoy our good fortune by going for a stroll. We take the opportunity to wine and dine well. Why not? Our printing press is rolling!

Two hours later we pass by the table to collect our winnings. During this time, the coin has been tossed one hundred times. How much money will we have won? A few thousand dollars, we anticipate.

So we are astounded when the croupier only hands us 36 dollars. And people who stayed at the table assure us that, oddly enough, heads and tails each came up exactly 50 times. The chances were exactly balanced. So how could we have lost money?

Very simply! As the game was played sequentially, the result is the product of 100 dollars multiplied 50 times by 140% and another 50 times by 70%. Calculate the result – it is 36 dollars. The fact is that 140% times 70% is 98%, which means that we were losing 2% of the money on the table for each heads and tails sequence.

This surprising result, which you can only obtain once you have done your sums carefully, is a curiosity arising from a field known as recreational mathematics, which is particularly useful for its applications in other areas. The lesson? It is one thing to add up your expected winnings, it is quite another matter to multiply them.

Take another example regarding the calculation of bank fees. If a bank makes you a loan, charging fees of 4% per quarter but paying 8% interest per annum on your long-term deposits, who ends up winning?

MONSIEUR BERTRAND

We expect to receive two Olympic medals, and we know that neither of them is bronze. There are three boxes in front of us, each containing two medals. One contains two gold medals (GG), another two silver medals (SS), and the third one gold and one silver medal (GS). The boxes are indistinguishable from one another, each with two drawers containing one medal. This is all the information we know. We select a box at random, open one of the drawers and find a silver medal inside. What is the probability that there will be a gold medal in the other drawer of this box? That seems easy. We have eliminated the possibility of the box containing two gold medals (GG), so there are two hypotheses: we selected the box with two silver medals (SS) or the box with one gold and one silver medal (GS). That seems to be it – the probability of finding gold in the other drawer of the box is 1 in 2.

Or is it? Let's look at the problem from a different perspective. The way the problem is set, the probability of choosing any drawer would be the same. This means that we could have selected a drawer in the box with two silver medals (SS), the other drawer in the same box (SS), or the drawer with the silver medal in the mixed box (GS). So we have three hypotheses. We find gold in the other drawer in only one of those three hypotheses. Which means that the probability of finding gold after silver is only 1 in 3.

If this has confused you, don't despair. Just think a little and check that the second version is right. This puzzle was invented 110 years ago by the French mathematician Joseph Bertrand and many people still find

N. Crato, *Figuring It Out*, DOI 10.1007/978-3-642-04833-3_53,
© Springer-Verlag Berlin Heidelberg 2010

it confusing. It has nothing to do with the Olympics, where little is left to chance.

J.L.F. Bertrand

Joseph Louis François Bertrand (March 11, 1822 – April 5, 1900)

BOY OR GIRL?

Mary and John have two children. Their first-born is a boy named Jack. What is the probability that the couple have two children of different genders? This seems at first to be a ridiculously simple question. If we concede that it is just as probable for a boy or a girl to be born, and if we also concede that this has nothing to do with the gender of the first-born baby, then there is no doubt that the probability that the second child will be a girl (therefore not the same gender as the first-born) is 1 in 2. And that is the answer: 1 in 2.

Simple, right? But now let us consider another couple, Josephine and Joseph, who also have two and only two children. We know that one of them is a boy, but we don't know if he is their oldest or youngest child. The question is: what is the probability that this couple has a boy and a girl? Lulled by our previous findings, we predict that their chance is also 1 in 2. And that's that. Except that, surprisingly, this answer is wrong. The right answer is 2 in 3.

In actual fact, we know that Josephine and Joseph could have two boys (BB), two girls (GG), a boy followed by a girl (BG) or a girl followed by a boy (GB). The only thing we know for sure is that one of their two children is a boy, so the two girls (GG) hypothesis is ruled out. Of the three remaining hypotheses (all of which are equally probable) two (BG and GB) involve children of different genders. Therefore the probability is 2 in 3. Hard to believe, isn't it? That is why the paradox is so entertaining. We have to think a little to understand it.

N. Crato, *Figuring It Out*, DOI 10.1007/978-3-642-04833-3_54,
© Springer-Verlag Berlin Heidelberg 2010

A PUZZLE FOR CHRISTMAS

Everyone knows that Santa Claus likes to please people. But he doesn't like to waste presents. He put money in my stocking. But we came to an agreement, he and I. Or rather, he explained the rules of the game to me.

He appeared at my house after I had spent an evening out with friends. He said: "My dear Nuno, you have worked hard this year, and I want to reward you. But as mathematicians think they know everything, I am going to give you a lesson in humility, with a lot of cash on the side."

I rubbed my eyes, dumbfounded (maybe I had drunk more wine at dinner than I'd realized). I looked up again, but the old man with the white beard was still there.

"Hang two stockings beside the chimney" he instructed. "I'll put a thousand dollars in the stocking on the left, but I'm not sure about the other stocking; I'll leave either a million dollars or nothing in it. You'll have two choices: you can either choose the stocking on the right and throw away the stocking on the left, or you can choose both stockings. If you choose the stocking on the right, you will find a million dollars in it. If you choose both stockings, there will be nothing in the one on the right, so you'll just have a thousand dollars."

I was confused: "Santa Claus! I guess I understand the rules, but there's one thing I don't get. You're going to put the money in the stockings before I decide which to choose. How will you know how much money to put in the stocking on the right if you don't know what I'm going to do?"

N. Crato, *Figuring It Out*, DOI 10.1007/978-3-642-04833-3_55,
© Springer-Verlag Berlin Heidelberg 2010

"Well, that's where you're wrong. I have a tremendous capacity for telling the future. I am almost absolutely certain that I know what you'll choose. I've already played this game with people who are much cleverer than you, and I've always guessed right. I have even played this game with Archimedes, Al-Khwarizmi, Pedro Nunes, and other mathematical geniuses, except it wasn't in dollars because they used other currencies in those days."

I thought and thought about it before I fell asleep. The old guy with the white beard didn't turn up again. On Christmas Eve I hung my two stockings beside the chimney. I went to bed happy, thinking I had found the answer. It was so simple. I just had to choose the second stocking, the one on the right, with the million dollars in it. How lucky I was! The old guy would play along!

The following morning the two stockings bulged. Santa Claus had kept his word. Had he doubted that I was going to choose the stocking on the right and throw away the one on the left?

Just at that moment, however, something stopped me. I've always been a bit of a cheapskate and now I couldn't bring myself to throw away the stocking on the left that was surely stuffed with 1000 dollars. What if I were to just take both stockings?

I was absorbed in this internal debate when I heard Santa Claus's voice in my head telling me "I've already played this game with people who are much cleverer than you"... And then I reminded myself, he could predict the future. If I chose both the stockings, he would have foreseen that and would not have left a million dollars in the stocking on the right. It was better not to be greedy and to just throw away the stocking on the left.

But then a thunderbolt hit me! The stockings were there in front of me with the money inside. The one on the left definitely contained a thousand dollars. The one on the right either had a million or zilch. It was best to grab them both.

I couldn't make up my mind. If the most logical choice was to take both, Santa Claus would have known that this is what I would do and he would have put nothing in the stocking on the right. But how could the logical choice leave me with only a thousand dollars when a million was a real possibility?

Christmas Day passed, the New Year came, and I still don't know what to do about the stockings. The old guy's thunderbolt really got me mixed up!

***

The Santa Claus in this story is actually a physicist named William Newcomb. His paradox involving a highly skilled *Predictor* was published in 1969 by the U.S. philosopher Robert Nozick (1938–2002), and immediately gave rise to a heated debate.[1] Logicians, mathematicians, economists and theologians have discussed possible solutions for this paradox, without coming to any consensus.

The problem seems to be related to the famous prisoner's dilemma, to the theory of free will when confronted by an omniscient being, as well as to the irreversibility of time, and other crucial questions in the fields of logic and philosophy. A paradox's possible ramifications, and the difficulties in solving them is what make them so fascinating. Don't worry if they stumped you too. Have a good one!

---

[1] R. Nozick, Newcomb's Problem and Two principles of Choice, in N. Rescher and S. Library, Editors, *Essays in Honor of Carl G. Hempel*, D. Reidel, Dordrecht, 1969, p. 115.

CRISIS TIME FOR EASTER EGGS

Every year the Easter Bunny has an infinite number of eggs available for distribution. Nobody knows where he gets them from, or how he manages to get them to all the children on Easter morning.

But this year the bunny was anxious, as he had heard talk of the economic crisis and thought this might affect the supply of eggs. If there wouldn't be enough eggs to go round, how could he fulfill his mission this year and the next? He decided the best thing would be to start saving eggs.

He made his preparations. He took his magic bag with room for an infinite number of eggs, and made up an identical bag. He cut a hole in the bottom of the first bag and told his assistant Tricksy to get into the bag and stay there to help. As the Easter Bunny and his assistant are infinitely quick and infinitely efficient, they can easily handle an infinite number of eggs in a short time.

Now, and this is a fact known only to a very few people, when the Easter Bunny receives Easter eggs they are already numbered: 1, 2, 3, ... The Easter Bunny decided to place the eggs in groups of ten into the first bag: first the eggs numbered from 1 to 10, then those numbered from 11 to 20, then those numbered from 21 to 30, and so on. His assistant, Tricksy, hiding in the bottom of the bag, was tasked with taking the first egg from each group of ten and placing it into the second bag. So when the group of eggs numbered from 1 to 10 were put in the bag, Tricksy removed egg number 1 and placed it in the second bag. When the group of eggs numbered from 11 to 20 arrived, Tricksy grabbed number 11 and put it in the second bag. And the two of them did this again and again ...

N. Crato, *Figuring It Out*, DOI 10.1007/978-3-642-04833-3_56,
© Springer-Verlag Berlin Heidelberg 2010

The Easter Bunny knew that by using his method they would end up with an infinite number of eggs in the first bag to be used this year at Easter, and with an infinite number of eggs also in the second bag, to be kept for the following year. There would be one egg in the second bag for every nine eggs in the first bag. But after repeating this process an infinite number of times there would be an infinite number of eggs in each bag. Infinity is like that, full of surprises. Although you might think that the second bag contained nine times fewer eggs than the first, actually they both contained an infinite number of eggs. So any problems with next year's egg supply that might be caused by the crisis had been solved.

But the Easter Bunny hadn't reckoned with the tricks perpetrated by his assistant Tricksy, who'd switched the numbers by sticking the number 11 label on egg number 2, the 21 label on egg number 3, and so on. This way, Tricksy grabbed egg number 1 when the Easter Bunny put the group numbered 1–10 in the bag, took out egg number 2 when the second group was placed in the bag, removed egg number 3 when the third group was put in, and so on.

But Easter Bunny, who is used to Tricksy's foolishness, didn't really mind. He kept on putting eggs into the first bag in groups of ten, while Tricksy continued to take one egg from each group and put it into the second bag, until their task was completed. However, that was when the Easter Bunny realized that there was nothing left in the first bag, all the eggs had ended up in the second bag. How could that be possible, when only one egg was placed in the second bag for every nine put into the first bag?

But it is possible! Think of any egg whatsoever; let's say number 27 or even number 10145? Well, these eggs disappeared from the first bag when the Easter Bunny put the corresponding groups of ten into the bag. If you think it through, this means that every single egg eventually ends up in the second bag.

Are you confused? So was the Easter Bunny! Although he thought there were nine times more eggs in the first bag than in the second, he realized that all the eggs were in the second bag. So this year's eggs will be from the stash intended for next year, and that is the crisis.

INDEX

N. Crato, *Figuring It Out*, DOI 10.1007/978-3-642-04833-3,
© Springer-Verlag Berlin Heidelberg 2010

Printed in the United States
By Bookmasters